MEAN SEASON

Florida's hurricanes of 2004

PIER, INTERRUPTED
Hurricane Frances ripped away sections of the municipal pier off Lake Worth Beach when it came ashore just after midnight Sept. 5. This photograph was taken a few hours later. The landmark structure has suffered storm damage and been rebuilt many times over the years.

CYDNEY SCOTT

Published by Longstreet Press

Copyright© 2004 The Palm Beach Post

Printed in the United States of America
1st printing 2004
Library of Congress Catalog Card Number: 2004115597
ISBN: 1-56352-745-6

This book was printed by R.R. Donnelley in Roanoke, Va.

NORTON FAMILY PHOTO

GREG LOVETT

On the cover
"MY HEART IS STILL THERE"
Until June, Alex Norton, 19, had spent her whole life in this waterfront home on Pensacola's Scenic Highway (**above**). Then her family sold it and moved to a different neighborhood, but Alex, a junior at the University of West Florida in Pensacola, would pass the house on her way to class. But when Hurricane Ivan struck the Panhandle on Sept. 16, storm surge churned into Escambia Bay, and Alex's childhood home was totally destroyed. Alex's mother, Sandy Norton, said Alex (shown in the cover photograph in what was once the home's doorway) and 14-year-old sister, Kristin, were still in shock. "We've been through Category 3 hurricanes before, with Erin and Opal, but nothing like this." Their new home was spared major damage, but all of Alex's memories were with the old one. "I miss my room," she says, "and the bookshelves my grandpa built." Now, as she passes the site, the tears inevitably come. "My mom keeps saying, 'Home is where the heart is,' but I think my heart is still there."

MEAN SEASON

Florida's hurricanes of 2004

EDITOR

Jan Tuckwood

PHOTO EDITOR

Mark Edelson

DESIGNER

Mark Buzek

HURRICANE HISTORY

Eliot Kleinberg

CAPTION WRITER/COPY EDITOR

Margaret McKenzie

PUBLISHING COORDINATOR

Lynn Kalber

––––––––––

PHOTOGRAPHY

The staff of The Palm Beach Post

Assistant managing editor/photo: Pete Cross
Deputy director of photography: John J. Lopinot

GRAPHICS

The staff of The Palm Beach Post

Art director: Christopher Smith

REPORTING

The staff of The Palm Beach Post

Managing editor: John Bartosek
Deputy managing editor: Bill Rose
Metro editor: Carolyn DiPaolo
State editor: Paul Blythe

SKY LIGHT
George Henderson sits and stares up at the sky through his missing roof, damaged by Hurricane Jeanne. He lives with his son at the Pleasure Cove mobile-home park in Fort Pierce.

GREG LOVETT

CONTENTS

August 14, 2004

September 17, 2004

September 6, 2004

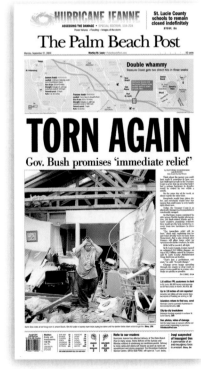

September 27, 2004

ALABAMA

GEORGIA

Mobile

Pensacola
Destin • Grayton Beach
Gulf Shores • Panama City
Panama City Beach

Tallahassee

FLORIDA

Jacksonville

St. Augustine

Daytona Beach

Orlando

Kissimmee

Clearwater • Tampa
St. Petersburg

Melbourne

Lake Wales

Vero Beach

Fort Pierce

Sarasota

Stuart

Port Charlotte
Punta Gorda

Lake O.

Jupiter

Belle Glade

West Palm Beach

Captiva • Fort Myers

Sanibel

Fort Lauderdale

The Everglades

Miami

Ivan

Jeanne

Frances

Charley

IVAN

Category: 3

Landfall: Sept. 16, Gulf Shores, Ala. Eastern part of the storm hit Florida's Panhandle

Top sustained wind speed at landfall: 130 mph

JEANNE

Category: 3

Landfall: Sept. 25, south end of Hutchinson Island

Top sustained wind speed at landfall: 120 mph

FRANCES

Category: 2

Landfall: Sept. 5, Sewall's Point

Top sustained wind speed at landfall: 105 mph

CHARLEY

Category: 4

Landfall: Aug. 13, Punta Gorda

Top sustained wind speed at landfall: 145 mph

SAFFIR-SIMPSON SCALE OF HURRICANE STRENGTH

Category 1: 74-95 mph

Category 2: 96-110 mph

Category 3: 111-130 mph

Category 4: 131-155 mph

Category 5: 155 mph and up

"One hurricane is enough for most people for a lifetime. It's going to be a year to tell your grandchildren about."

MAX MAYFIELD, director of the National Hurricane Center to emergency management officials in Indian River County

Graphic by BRENNAN KING

Four storms.
One mean season.

By
MICHAEL BROWNING
Palm Beach Post staff

Charley: 145 mph. Frances: 105 mph. Ivan: 130 mph. Jeanne: 120 mph.

These were the wind speeds at landfall in Florida belonging to four astonishing hurricanes that buzz-sawed the state, east coast and west coast, Panhandle and interior, stampeding people from Key West to Pensacola.

The storms moved like slow local trains on strange, self-chosen rails, obeying some terrible schedule, making all stops in between unknown destinations. They whistled through, in just a few weeks' time, blam, blam, blam, blam, solving huge subtraction problems with ruthless arithmetic.

Nothing in the state's meteorological history rivals this one-blamed-thing-after-another run of spectacular bad luck. Records kept since 1851 show that never has Florida been so particularly punished, so repeatedly, in such a short period.

"Ah, Gertrude, Gertrude, when misfortunes come, they come not single spies,

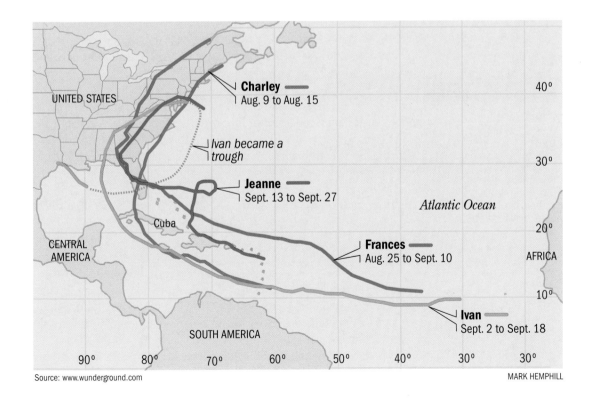

Source: www.wunderground.com

MARK HEMPHILL

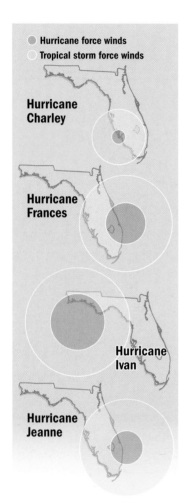

Hurricane force winds
Tropical storm force winds

Hurricane
Charley

Hurricane
Frances

Hurricane
Ivan

Hurricane
Jeanne

Source: National Hurricane Center
CHRISTOPHER SMITH, STEVE LOPEZ

COMPARING THE SIZE
OF THE STORMS
Charley was the most intense
of the four hurricanes but also
the smallest. Frances' winds
affected almost the entire state.
Ivan and Jeanne were also large.

but in battalions!" groans King Claudius in Shakespeare's *Hamlet*.

In one awful trice of time, little longer than it takes the moon to wax and wane and reappear again, the whole state became acquainted with wind-borne grief, fourfold. One storm, Ivan, apparently liked Florida so much it wandered up north and then looped back and paid us a second visit.

About 2,200 people, mostly poor Haitians and islanders, are dead now, who were alive before a trick of air and sea-warmth above sunny oceans off the coast of Africa formed and swirled into circles. Rain, mud, flooding, flying timbers and fallen power lines extinguished them. They drowned in wind, water and slime. The hurricanes vacuumed the breath from their lungs.

Florida's death toll is reckoned at about 125 for all four storms. The economic damage to Florida alone is reckoned at $29 billion to $41 billion: $14 billion for Charley; $4 billion for Frances; $5-$15 billion for Ivan; $6-8 billion for Jeanne.

A deceptive Charley

A dry report from the Centers for Disease Control in Atlanta summed up the strangeness of just one storm, Charley:

"As of September 1, a total of 31 deaths had been reported; 12 (39%) occurred on the first day of the storm, and eight (26%) additional deaths occurred during the next 2 days. Decedents ranged in age from 6 to 87 years (mean: 54 years; median: 56 years); 24 (77%) were male. Of the 31 deaths, 24 (77%) were classified as unintentional injury, six (19%) were attributable to natural causes, and one death was a suicide. Of the 24 unintentional deaths, 17 (71%) were trauma related, three were caused by carbon monoxide (CO) poisoning, and one each were caused by electrocution and drowning; two deaths involved at least two factors in combination (i.e., trauma and electrocution or CO poisoning and burn).

"Of the 18 deaths related to trauma and drowning, 11 (62%) occurred on the day the storm made landfall. Of those 11 deaths, nine were directly related to the storm, and two were indirectly related (i.e., an automobile crash in which a traffic light was out and a fall by an evacuee in a hotel room). Four trauma deaths resulted from motor-vehicle crashes, four from falls, three from implosion of a shelter (i.e., mobile home or shed), two from falling trees, two from flying debris, one from an uncertain cause, and one from a crush injury.

"Of the six deaths related to natural causes, four resulted from exacerbation of cardiac conditions and two from exacerbation of preexisting pulmonary conditions. Two persons lost power during the storm and did not have access to their oxygen. One man likely had a heart attack during cleanup. Three men died of heart failure, one during the storm and two in the days after the storm. Of these three deaths, two were associated with exposure to extreme heat.

"The suicide death involved a man who became despondent after losing his home and possessions to Hurricane Charley; his death resulted from a witnessed self-inflicted gunshot wound to the head."

And that was just the first storm. The Morbidity and Mortality reports aren't in yet for Frances, Ivan and Jeanne.

Charley was perhaps the worst of the four, in that it was so deceptive. It quickened from 110 mph to 145-180 mph in just three hours, then punched in southward of where it was supposed to hit, taking thousands by surprise.

Charley seemed destined to strike Tampa and St. Petersburg. Instead, it hit Fort Myers and Sanibel Island. It tore the coast to pieces, bulldozed homes and trailer parks, toppled oaks as far inland as Arcadia, ripped through sandbars and beaches like a cosmic Ditch Witch.

Above all, Charley knocked the state dizzy in terms of rescue and relief.

But Charley was just the bitter aperitif. A full three-course dinner came afterward, unordered, still being digested and paid for.

Slow-moving Frances

Next, Frances. Just as you remember your first kiss, people who had never seen a hurricane before will always remember Frances.

At 11 p.m., Sept. 4, 2004, Hurricane Frances' western eyewall hit Sewall's Point, Stuart, Jensen Beach and Port Salerno. It moved at 5 mph, as slowly as a drunken elephant, with an eye 50 miles wide. This meant some areas had to spend more than 24 hours under the pitiless flail of Frances' inexhaustible winds. It was a Category 2 storm — which was a mercy, in that it was supposed to strengthen to Category 3 and didn't. It was the size of Texas. It made its way all the way up to Quebec before dissipating.

We saw it first, and it lessened us considerably. Frances sheared away trees and thinned out the mobile-home population.

Power lines sagged, bellied and broke. Transformers exploded in beautiful blue-white arcs of light. A rubber doll's head washed into my lawn hedge, along with a fish. Some 3 million cubic yards of tree limbs and leaves all disbranched and flew away from their trunks, a volume of vegetation 2.18 times the size of New York's Empire State Building. Some of it has yet to be gathered up.

Power outages affected up to 6 million people. In the powerless aftermath, ice and gasoline were sought after like spices from Arabia. Traffic lights were out everywhere, and cars crawled through intersections, usually courteously, at horse-and-buggy speeds.

Neighbors became more neighborly. Misery loved company. Cold chili out of a can kept body and soul together. "Simplify, simplify!" Henry David Thoreau once urged Americans. We simplified.

But there were also acts of random cruelty and greed. About 20 looters were arrested statewide in Frances' wake, 10 of them in Palm Beach County. About 150 extraordinarily despicable pet owners showed their true colors, abandoning their animals to be euthanized at the county's Animal Care and Control center, before turning tail and scampering to safety themselves.

"It happened," said center director Dianne Sauvé. "I can't fathom how anyone can leave an animal behind at a time like this, but it happened. We begged them not to do it, but they did it anyway. The staff was crying. I have been working here two years, and several times I had to go into my office and sit alone, to compose myself, after what I had seen.

"The sad thing is, we don't have any animal-friendly hurricane shelters in the state. We were at capacity. But we are an open-door facility, the largest of our kind in the state. By law, we cannot turn anyone away. We told them what was going to happen to their pets. We begged them not to leave them. But they did."

Double-dipping Ivan

Hurricane Ivan began as Tropical Depression Nine, 555 miles southwest of the Cape Verde Islands, on Sept. 2. By Sept. 9, Ivan was a Category 5 hurricane, with sustained winds of 160 mph.

It skated past Jamaica, brushed the western tip of Cuba, passed through the Yucatan channel and the Gulf of Mexico, weakening only to 140 mph. It made landfall at Gulf Shores, Ala., at 2 a.m. on Sept. 16, with 130-mph winds. It severed the Interstate 10 bridge over Pensacola Bay.

Shameful to relate, we in Palm Beach County regarded Ivan's trajectory with relief and pleasure. "Our thoughts and prayers," as politicians love to say, went out to the poor Panhandle people in its path, but we gave at the office. We'd already had our hurricane. It was plenty and to spare for us. Meanwhile, Ivan's victims had our full moral support and deepest, deepest sympathy.

Ivan shambled on up into Virginia like a wet "Wish You Weren't Here" postcard

Neighbors became more neighborly. Misery loved company.

JENNI GIRTMAN

COVERING THEIR BASES
The plywood shutters at Skeffington's Furniture in Stuart got a workout.

> *"Frances was the reaper. Jeanne was the gleaner."*

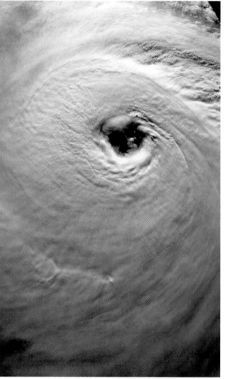

EDWARD M. FINCKE/NASA

EYE OF IVAN
The eye of Hurricane Ivan, entering the Gulf of Mexico with sustained winds of 160 mph, can be seen from the International Space Station about 230 miles above the Earth, on Sept. 13.

from Florida before losing its tropical storm status. It drifted into Canada, combined with a local storm, and knocked down power lines in Nova Scotia.

Then, Ivan did something truly astounding.

On Sept. 20, a kind of forgotten caboose of Ivan combined with a small surface low-pressure area to reform into another storm. This one looped back across Florida, moved into the Gulf of Mexico and made landfall at Cameron, La., on Sept. 23. It wobbled over into Texas and died.

Ivan caused immense flooding. Condominiums on Perdido Key, near Pensacola, were leveled. The Chattahoochee River rose to levels unseen in 100 years. The Blue Ridge Parkway in North Carolina sustained major damage, and rivers in western North Carolina overflowed their banks. Ivan destroyed seven oil platforms in the Gulf of Mexico and spawned tornadoes as far north as Maryland.

What would have befallen us, had Ivan come here, does not bear thinking about.

Copycat Jeanne

Jeanne was the 10th named storm, the fifth major hurricane since the 2004 season began with Alex.

Jeanne took a shine to southeastern Florida right from the start. It followed Frances with uncanny exactitude, coming ashore at 11:50 p.m., Sept. 25, at Hutchinson Island, just east of Stuart, three weeks to the day after Frances had piled into the same place.

It was a faster, sharper, meaner storm than Frances, a Category 3, not 2, with winds of up to 120 mph. Once again, mobile homes flinched and fled into the sky. Once again, solid concrete buildings groaned and shuddered. Once again, trees, fences and roofs took wing. Once again, boats became submarines.

Mailboxes danced swift, swooning tangos, while traffic signs performed frantic semaphore wigwags, fanned by furious winds. Lake Okeechobee turned brown, its shallow bottom puddinged up by titanic eggbeater gales. The Hoover Dike surrounding it held firm.

Frances was the reaper. Jeanne was the gleaner. Palm Beach, Martin and St. Lucie counties suffered a double diminishment. Haiti saw the worst of it. Even though Jeanne was only a tropical storm when it crossed the poorest nation in the Western Hemisphere, it dumped 13 inches of rain on the place and managed to kill nearly 2,000 people in mudslides from deforested hilltops.

Jeanne wandered up into New Jersey, causing flash floods in Trenton and Philadelphia. Tornadoes touched down in Wilmington, Del., and Cherry Hill, N.J.

The door ajar

On we go, betwixt wind and wave, palm frond and sandpile, in our hyphenated, denuded paradise, stuttered over by storms.

The October sun seems bright, and the skies are swept by crisp breezes. You would never dream what a dank, dreadful churning we've gone through, what wild nights we've seen, what tumults in the dark have rumbled our lawns and roofs, how frail and insubstantial our little bungalows are.

The Igloo cooler goes back into the garage, along with the generator.

The flashlights get put away until next year.

The Chef Boyardee ravioli can keep indefinitely.

There are enough leftover batteries to light a small city.

Have you met our two twins, Flotsam and Jetsam?

For any of us endowed with a memory of 2004, the September sky will always seem vaguely ajar, a bright, heavenly unhasped door not quite shut. We know not when it will swing full open again, nor what may come through it.

Something; inevitably something.

THE GET-READY-THERE'S-ANOTHER-HURRICANE-COMING-THOUSAND-YARD STARE

DON WRIGHT

Lessons learned from four hurricanes

Natural disaster sparks unnaturally high anxiety:

"I knew people wanted water and canned foods . . .
but I didn't know doughnuts were on that list of supplies.
But that's why they call it comfort food;
they want to feel good after something bad happens."

NAT SIEGEL,
vice president of Krispy Kreme in South Florida

"Sometimes, it feels like this is a test of
resiliency for our state.
Other times, I feel like I'm Bill Murray
in 'Groundhog Day.' "

GOV. JEB BUSH
as Hurricane Jeanne approached Florida

"The more people are focused on The Weather Channel,
the more stress."

STEVE MOORE,
chaplain at St. Lucie Medical Center

"We've always said: Holidays and
hurricanes are our busiest times."

BARBARA WHEELER
at King's Liquors Lounge and Tobacco Land on
Lake Worth Road, where business was brisk.

There's nothing like a cup of hot coffee and a hamburger:

"If I went one more day without a cup of coffee,
my husband was going to throw me out."

WOMAN WAITING AT STARBUCKS
in the Mall at Wellington Green a few days after Frances hit.

"Tastes like prime rib and
Dom Perignon."

JEAN MARTIN
of West Palm Beach, as she ate her first McDonald's
double-cheeseburger and Diet Coke after Frances.

ALLEN EYESTONE

FIRST DISASTER . . . BUT NOT THE LAST

Mobile homes and cars in Port Charlotte (**above**) lie tossed and shredded after Charley came ashore on Aug. 13. Devastation from the small but powerful hurricane shocked Floridians, who hadn't suffered a direct hit from a major hurricane in years. Little did we know three more would be headed our way over the next six terrible weeks.

ALLEN EYESTONE

A NEW BREAK IN NORTH CAPTIVA ISLAND

Before Charley, North Captiva Island was one barrier island, with trees and sand filling in the blank spot shown here. The storm surge created a new inlet. The dark spots in the water are pieces of trees blown apart by the storm.

HURRICANE CHARLEY

- **Category 4 hurricane**

- **Landfall:**
  Aug. 13, Punta Gorda

- **Top sustained wind speed at landfall:**
  145 mph

- **Size of eye:** 10 miles

- **Notable characteristics:**
  A fast-moving and narrow storm, its hurricane-force winds might have been only a few dozen miles across. It tore across the peninsula in hours, causing extensive damage in Orlando and even the Daytona Beach area.

- **Deaths**: 33

- **Damage:**
  Up to $14 billion

- **Where deaths occurred**

Brevard	1
Charlotte	5
Collier	3
DeSoto	1
Hardee	2
Highlands	1
Lee	6
Orange	3
Osceola	1
Polk	6
Sarasota	2
Volusia	2

Source: Florida Division of Emergency Management, Oct. 19, 2004

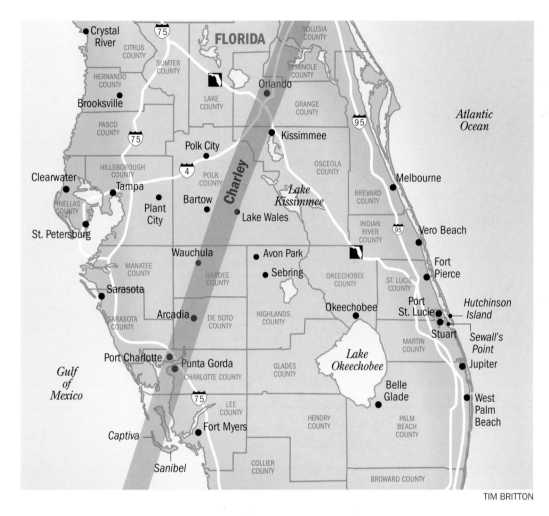

TIM BRITTON

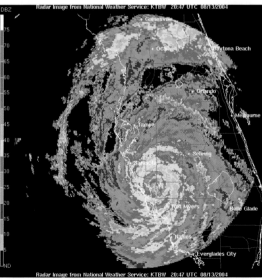

Charley buzzed through Florida quickly, but it took southwest Florida residents by surprise when it made a sudden easterly turn.

Friday the 13th: Charley's radar image at 4:47 p.m. on Aug. 13.

TWISTER
A waterspout forms off of Fort Myers on Aug. 12, the day before Hurricane Charley struck Florida's west coast.

Hurricane Charley

The hotel owner told us to leave.
"He really, really didn't want us to die in his hotel. . ."

WINDS OF CHANGE Robert Conidaris bragged about the stoutness of his 25-year-old hotel, the Lani Kai Beachfront in Fort Myers Beach. After Charley became a Category 4, however, he ordered *The Post's* news team to leave "now, for your own safety."

Caught in the storm

As Charley roars in, confidence turns to desperation

Story by
KIMBERLY MILLER
Palm Beach Post staff

Photography by
GARY CORONADO
Palm Beach Post staff

We had less than 30 minutes to make a potentially life-or-death decision. It seemed like 30 seconds, and for a few terrifying moments, we thought we had chosen wrong.

Palm Beach Post photographer Gary Coronado and I had driven to Fort Myers Beach to cover Hurricane Charley. We wanted to talk to people who had defied a mandatory evacuation order of Lee County's barrier islands before we headed back to the mainland over the high, arching Estero Bay Bridge. We wanted to stay until the last minute to soak up as much of Charley as possible without actually getting stuck in it.

That was the plan, anyway.

In no time, Hurricane Charley's screeching winds ramped up from a Category 2 to a 3, and then to what few predicted, expected or prepared for — a devastating

SURF'S UP AT DIAMONDHEAD
Storm surge from Hurricane
Charley (**above**) crashes
against the outdoor bar of the
DiamondHead Beach Resort,
where, trapped on Estero Island,
The Post's Kimberly Miller and
Gary Coronado found refuge as
Charley made landfall.

RESERVED PARKING
A convenience store in
Fort Myers Beach (**right**) plays
host to two unlikely visitors, jet
skis propelled there by waves
whipped by Hurricane Charley's
145-mph winds.

Category 4, with winds swirling up to 145 mph.

Through the next 12 hours on the island, I cried only once.

At first, the island was remarkably peaceful. Most people on the beach and by the pier were joking, and nearly all of them were drinking beer — even at 9 a.m. — and I thought, "Man, if this thing hits hard, there's going to be a whole lot of drunk people trying to deal with it."

Actually, I wasn't scared. After being psyched out by years of shrill, unrequited hurricane warnings, I had adopted the "it's not coming here" attitude.

We wandered into the Lani Kai Hotel, looking for a place to transmit our stories and pictures to *The Post*. After the hotel owner boasted of how his 25-year-old structure could withstand 125-mph winds, we started to feel pretty safe, and started thinking we might just stay on the island.

By the time TV news reported that Charley was a Category 4 and headed straight for Lee County, the Lani Kai owner became a little more concerned. He said we should leave "now, for your own safety." I got the impression he really, really didn't want us to die in his hotel.

With nowhere to stay, we decided to leave. But then we spotted a couple hauling two pet carriers down the street and leaning into a wind blowing hard enough to take a grown man off his feet. I pulled over for Gary to get the shot. It was pouring. The couple was struggling in the wind when one of the pet carriers fell apart and all of their stuff started blowing down the street.

Instinctively, we decided to help the couple chase after their belongings. I pulled the truck up and started frantically throwing stuff out of our back seat so they could jump in. We dropped them down the street at a friend's house.

It was at that point, I think, that Gary and I realized it was too late too leave. We had missed our window of opportunity to get over the bridge safely.

Now I was scared.

I spun the truck around, and we headed for the DiamondHead resort — a newer hotel that had cars in its parking lot. A sign of life.

We ran around the hotel with winds gusting and desperately pulled on the hotel doors. I don't remember thinking much at the time, except maybe that if we couldn't get into this hotel, I would go back to the nearby house and ask for help.

Finally, a door to a stairwell gave, and on the second floor, thankfully, we found about 60 people hunkered in the hotel's ballroom. We could stay, the manager said, if we signed a waiver saying we wouldn't blame the hotel if we were hurt.

For the next few hours, as the storm battered the island, we watched from hurricane-proof windows the destruction outside. Wind whistled shrilly up the elevator shaft. One little girl kept saying, "I can't stand that noise, I can't stand that noise."

We ventured out once, but the wind was blowing the rain so hard it stung our faces. Around the pool area, the concrete pavers were being washed away by the waves and the pool water was almost indistinguishable from the Gulf.

It was hard to walk, so I ducked behind a concrete wall to avoid flying away. Power lines were dangling like Christmas-tree tinsel. We headed back inside.

But not before seeing something incongruous — a man in shorts and a tank top meandering down the street in the painful rain. How was he standing up? Was he drunk? We didn't stick around to find out.

Palm Beach Post reporter Kimberly Miller (**top**) and photographer Gary Coronado found themselves caught in the middle of Hurricane Charley. By the time they found shelter, Charley's winds were strong enough to knock a man off his feet.

Cheryl Prather of Fort Myers Beach (**above**, with her cat, Lestat) watches in fear as the vacant house next door caught on fire shortly after Charley passed through. People heard explosions at the home, where the fire raged for an hour before firetrucks were finally allowed back onto the island. At **right:** A renter bows his head as the fire approaches his apartment.

At about 6:30 p.m., Charley was far enough away that we could go out safely. We stepped out into a smoke-filled sky and ran to a house engulfed in flames with mini-explosions going off inside. Debris was everywhere. People, mostly in flip-flops and shorts, were surveying the damage.

The storm left full-sized sofa beds, green trash Dumpsters and dishwashers in the middle of Estero Boulevard, the main thoroughfare of Fort Myers Beach. At least one house was leveled. Part of a roof lay in the sand. Jet skis landed oddly in front of storefronts, trees were uprooted and stop signs were pulled out of their concrete anchors.

Probably the saddest disaster was the raging fire. It had started in an evacuated house but left neighbors Michael and Cheryl Prather shivering in the rain, afraid their own home would go up in flames.

Flames from the burning house lapped at their roof. They knew no help was coming. The fire department had evacuated at 1 p.m., and the fire raged for an hour before the firetrucks finally ventured back on the island.

"I actually thought it was safe now," Michael Prather said, sitting on the curb across from his small home. "Do you know if there are any firetrucks left?"

Other residents described "a wall of water" that hit when Charley pushed the shallow Gulf waters ashore.

"I'll never stay again," said a man who calls himself Red Man. "It was too scary."

Many residents said they weren't expecting a Category 4 and may not have stayed if they had known.

We turned down one street and into the path of a truck surrounded by people walking alongside it. I backed our car out of the truck's way, and we saw why people were walking with the truck.

A large manatee had been washed out of the bay, and people had rolled it onto a makeshift wooden stretcher towed behind the truck. But the stretcher was so waterlogged and rotten, it kept snapping, and the manatee would fall to the street with a thud.

People chanted "Charley, Charley," the name they had given the manatee.

That's when I finally cried. I don't know whether it was the sight of the helpless manatee or just a release of the stress of the day, but it didn't last long.

I took out my notebook again. Gary snapped more photos.

We learned later that Charley the Manatee made it back into the bay safely and swam away.

RAGTAG RESCUE
A manatee found stranded on Pearl Street gets pulled behind a truck on a makeshift sled by a band of Fort Myers Beach residents. The animal was released safely into the bay.

INDOMITABLE SURVIVOR
Everett Cowan, 91, experienced the eye of Hurricane Charley at Biel's Mobile Home Park in Punta Gorda. Although a tree crashed into his trailer, he emerged without a scratch. The elderly residents of Charlotte County, which has the nation's largest percentage of seniors, proved themselves to be a hardy bunch.

RICHARD GRAULICH

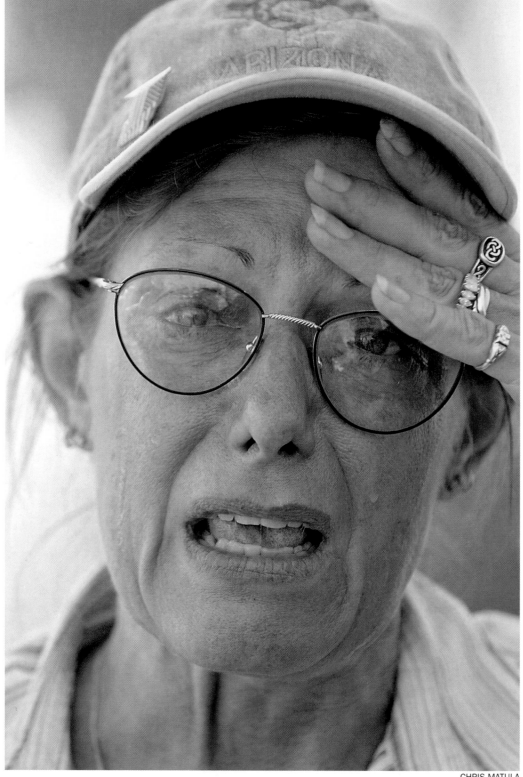

CHRIS MATULA

TOSSED LIKE TOYS
Charley's wind and storm surge – up to 15 feet – pushed boats
(**right**) onto land on the north end of Pine Island. The surge
also pushed so much sand over Sanibel Island that all the
shells were covered. Still, the storm surge could have been
much worse. Charley's power was on its right side, but it moved
quickly and hit the coast at an oblique angle.

ALLEN EYESTONE

GRANDMOTHER'S LEGACY

With his grandmother, 78-year-old Patricia Martin, safely in New York, Timothy Livingstone of Virginia Beach (**above**) searches for something to salvage in what's left of her Pine Island home on Aug. 15. His mother, Laurel Livingston, had recovered her father's ashes the day before Charley made landfall at the island northwest of Fort Myers.

ALLEN EYESTONE

UPROOTED

A boy ditches his bike to examine a tree uprooted in historic downtown Arcadia, in DeSoto County, which Charley ripped through on Friday, Aug. 13. Great, old trees suffered throughout Florida. In Lake Wales, Bok Tower Sanctuary's 245-acre historic landscape garden was all but destroyed, and a 100-plus-year-old oak tree, largest in the garden, was uprooted. A tree that President Calvin Coolidge and Edward Bok planted at the sanctuary's 1929 dedication ceremony still stands. At Orlando's Harry P. Leu Gardens, a 50-acre botanical garden, more than 100 trees were uprooted. The camellia collection, largest outside California, might not recover.

ARMED GUARD
As night falls on Aug. 13, Jane Duke (**above**) stands watch with her pistol as she salvages the remnants of her store, Westchester Gold Fine Jewelry & Antiques, on Tamiami Trail in Port Charlotte.

SAND TRAPS
Sand inundates the golf course at South Seas Plantation on Captiva Island (**left**). South Seas Plantation resort suffered enough damage to close most of the resort until Christmas 2004, with the golf course closed into 2005.

CHRIS MATULA

UNFATHOMABLE WRATH

Charley remained a hurricane throughout its trek across Florida to the Atlantic Ocean, killing 33 and leaving more than 1.3 million people without power. A homeowner (**above**) seeks assistance from a higher authority than FEMA. At **right**: The Lake Wales First Baptist Church lost its steeple to Charley.

LANNIS WATERS

CHURCH RUINED, BUT "NO FEAR"

The Rev. Jerry Kaywell (**above, center**) and Kathy Frey comfort Kathy's husband, Terry, who broke down in tears after touring the ruins of Punta Gorda's Sacred Heart Catholic Church. Father Kaywell said Mass on Aug. 15 by flashlight at the parish hall across the street. Under his robe, he wore jeans and a white T-shirt that said "No Fear."

"Our hope in sorrow and in woe..."

Two days after Charley visited, nearly 50 people found their way to Sacred Heart Catholic Church in Punta Gorda for the 7 a.m. service. By 8:30 a.m., there were 100. The main church building was destroyed, so they gathered in the parish center across the street. They had no power, so they shined flashlights. They sang one hymn they knew by heart – *Salve Regina.*

"Our life, our sweetness here below . . .Our hope in sorrow and in woe . . ."

Father Jerry Kaywell, who grew up in West Palm Beach, was amazed by the little things that remained at Sacred Heart. His vegetable garden was destroyed, but one pot of basil was untouched. Stained glass was blown out along one wall of the church. Yet, inside the tabernacle, he found a candle still burning.

"Triumph all ye cherubim . . . Sing with us ye seraphim . . . Heaven and earth resound the hymn . . . Salve, salve, salve, Regina . . ."

– KATHLEEN CHAPMAN

NOTHING STOPPED CHARLEY
Everyone in the aptly named Windmill Village mobile home park in Punta Gorda had a damaged or destroyed home. But everyone survived.

RICHARD GRAULICH

CHRIS MATULA

NICE DAY FOR A DIP
Joseph Barbee, 6, cools off in his grandparents' pool on Goldstein Street in Punta Gorda. Joseph's home near the Charlotte County Airport was destroyed by the hurricane, so his mother, Lavina Cissell, brought him to her parents' house, which her father remodeled two years ago to double the standards of the Miami-Dade building code. "We're lucky to be alive," she said.

HEADING . . . HOME?
Captiva residents finally got the go-ahead to return to their properties on Aug. 18, five days after Hurricane Charley ravaged the island. One man said his heart was in his throat when he crossed the bridge. "You prepare for the worst," he said. "But it's not as bad as it looks." The people of Palm Beach County and the Treasure Coast sent relief workers and supplies. Two weeks later, the situation would be reversed.

GARY CORONADO

GOD, BE MERCIFUL . . .
Flor Loperena (from left), Gladys Borin and Gladys' daughter, Nicole, pray at Friday morning mass at St. Ignatius Loyola Cathedral in Palm Beach Gardens on Sept. 3. Palm Beach County was under a hurricane warning as huge Hurricane Frances proceeded on a wobbly path toward the Florida coast.

SHANNON O'BRIEN

HURRICANE
FRANCES

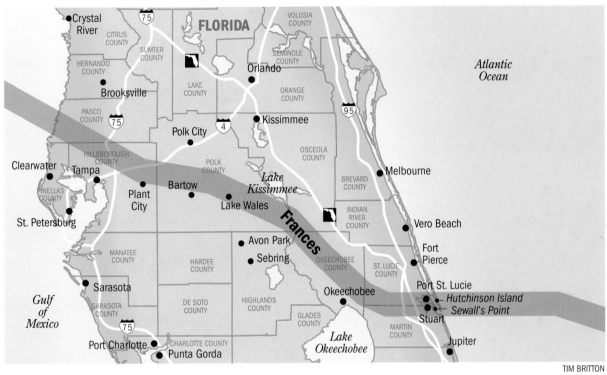

Crystal River · Citrus County · Sumter County · Hernando County · FLORIDA · Volusia County · Seminole County · Orlando · Orange County · Atlantic Ocean · Brooksville · Lake County · Pasco County · Kissimmee · Polk City · Osceola County · Clearwater · Tampa · Hillsborough County · Polk County · Lake Kissimmee · Brevard County · Melbourne · Plant City · Bartow · Lake Wales · Indian River County · Pinellas County · St. Petersburg · Frances · Vero Beach · Avon Park · Fort Pierce · Sarasota · Manatee County · Hardee County · Okeechobee County · St. Lucie County · Port St. Lucie · Sarasota County · Gulf of Mexico · De Soto County · Highlands County · Okeechobee · Hutchinson Island · Glades County · Sewall's Point · Stuart · Jupiter · Port Charlotte · Charlotte County · Punta Gorda · Martin County · Lake Okeechobee

TIM BRITTON

Frances was the first hurricane to strike Palm Beach County and the Treasure Coast in 25 years. Hurricane David hit as a Category 1 over Labor Day weekend in 1979.

- **Category 2 hurricane**

- **Landfall:**
  Sept. 5 (with gusting winds felt as early as the evening of Sept. 3), Sewall's Point

- **Top sustained wind speed at landfall:** 105 mph

- **Size of eye:** 50 to 70 miles

- **Notable characteristics:**
  A giant storm, some 400 miles across, that trudged across nearly the entire state over three days, bringing high winds and copious rain.

- **Deaths:** 37

- **Damage:**
  $4 billion to $8 billion in insured losses

- **Where deaths occurred**

Alachua	3
Broward	1
Collier	1
Flagler	1
Highlands	1
Hillsborough	2
Indian River	1
Lake	1
Lee	1
Marion	3
Martin	1
Orange	1
Osceola	2
Palm Beach	6
Polk	4
Putnam	1
Seminole	1
St. Lucie	2
Volusia	4

Source: Florida Division of Emergency Management, Oct. 19, 2004

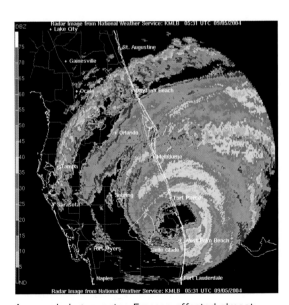

A meandering monster, Frances affected almost the entire state for almost 30 hours. This shot was taken at 1:31 a.m. on Sept. 5.

"The plywood ends at that guy in the white hat."

EMPLOYEE
at the Boynton Beach Home Depot

ALLEN EYESTONE

WHO WANTS PLYWOOD? ME! ME! ME!
Customers raise their hands for plywood outside Lowe's in Royal Palm Beach on Friday, Sept. 3, hours before Frances's winds picked up. By 11 a.m., the home-improvement giant's plywood supply was gone. At Home Depot, customers waited three hours to buy plywood. (Patience was a virtue before Frances hit. With 250 million Floridians ordered to evacuate, highways were jammed. It took 14 hours to drive from West Palm Beach to Georgia.)

CHRIS MATULA

GRAB AND GO
When Kodiak Construction in suburban West Palm Beach announced it would give away $30,000 of scrap plywood, the construction lanes of Southern Boulevard became a parking lot. Florida Highway Patrol Sgt. T. Kelly helps control the crowds, telling folks to pick up what they have and leave. The wood was gone in two hours, but not before two scavengers got in a fight.

BOB SHANLEY

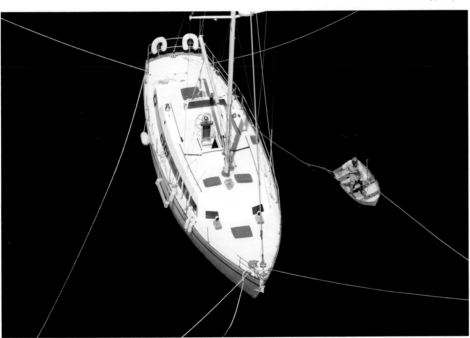

RICHARD GRAULICH

IT'S A ZOO IN THERE
To shelter them from Frances, spoonbills and ibises were moved to a women's restroom at the Palm Beach Zoo at Dreher Park. A Cooper's hawk, a baby ibis and several ducks died during the storm, probably of stress, said Keith Lovett, director of the zoo's living collections. Hurricane Jeanne also took a toll, killing a pair of toucans and several chickens.

DAMON HIGGINS

BILL INGRAM

Lewis, a 2-year-old North American black bear, gets a ride to a secure area of the Palm Beach Zoo at Dreher Park. The swirling storm damaged landscaping, enclosures and buildings at the zoo.

Did animals sense the storms were coming?

We humans had Doppler radar and the Cone of Probability. Animals didn't – but they seemed to know when the hurricanes were coming.

"I really, really believe animals can tell that something is coming," said David Hitzig, executive director of the Busch Wildlife Sanctuary in Jupiter. "All of our animals acted differently."

Hitzig and his staff are feeding and treating as many as 200 wild critters that were injured or orphaned in the storms, in addition to the 300 to 400 animals that live at the 20-acre sanctuary permanently. Four animals at

Busch died during Frances, including a bald eagle, who probably died of stress. The sanctuary itself suffered about $250,000 in damage. Lion Country Safari and the Palm Beach Zoo had more than $1.5 million in damage and lost revenue.

"Animals in the wild are more in tune to external influences than people," Hitzig said, explaining why they may sense when storms are coming. "For many years, we needed to rely on our senses. Now we're not required to, so we don't take notice."

– LAUREN GOLD

CYDNEY SCOTT

DEFIANT MESSAGE
The Pottery Playhouse (**left**), one of many boarded-up shops along Atlantic Avenue in Delray Beach's downtown district, appears to stare down the impending storm.

SPECIAL-NEEDS SHELTER
Robert Nadaskay of Delray Beach inside the special-needs shelter at the Palm Beach County Fairgrounds on Sept. 2. The shelter had 480 beds. Thousands of people stayed in shelters for several days during Frances.

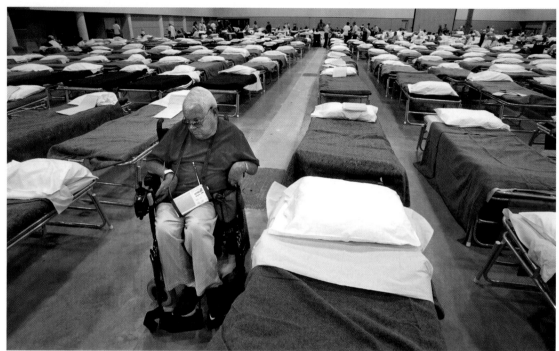

LANNIS WATERS

GARY CORONADO

NO HOME, NO PLAN
As the wind picks up Friday, Sept. 3, Betty Stiles, 51, stands outside the public library in downtown West Palm Beach, trying to figure out, along with a group of other homeless people, where they can ride out the storm.

UMA SANGHVI

MEGHAN McCARTHY

THE LONG WAIT
Inside the shelters, Frances was a bore rather than a fright. Martin Katz of Hutchinson Island (**above**) reads while lounging on school desks at Jensen Beach Elementary School on Sept. 3. At the Bear Lakes Middle School shelter in West Palm Beach, 5-year-old Cindy Gonzales (**left**) fell asleep by 9 p.m. on Sept. 3.

LIGHTS OUT
Police officers chat in the gym of Glades Central High School, one of two public shelters in Belle Glade, as lights are lowered at 10 p.m. Thursday, Sept. 2. People kept arriving at the school right up until 10:30, when doors were closed for the night. The Red Cross ran dozens of shelters in Palm Beach, Martin and St. Lucie counties during Frances, and the Palm Beach County chapter alone served 400,000 hot meals through Sept. 8 (that's nearly $700,000 in food).

The kindness of strangers:
Shelters provide security. . . and solace

Inside the shelters, personal dramas played out quietly. None was as poignant as Joane LiCalzi's last night with her sister, Joyce Haldeman of Port St. Lucie.

Joane, 68, came south from New Jersey to take care of Joyce, 69, who had battled pancreatic and lung cancer since March. When Frances threatened, the sisters sought shelter at the Port St. Lucie Civic Center.

There, in the waning winds of Hurricane Frances, Joyce died in Joane's arms.

"These people that I've never met before, they went out of their way to make sure she was comfortable," Joane said of the shelter volunteers.

One woman took the cross off her neck and laid it on Joyce.

Moments later, "my sister died very peacefully," Joane said.

"I'm not even a Floridian. But these people. . ." she said, her eyes filling with tears, ". . . these people lifted my heart."

— JOHN BISOGNANO

Joyce Haldeman

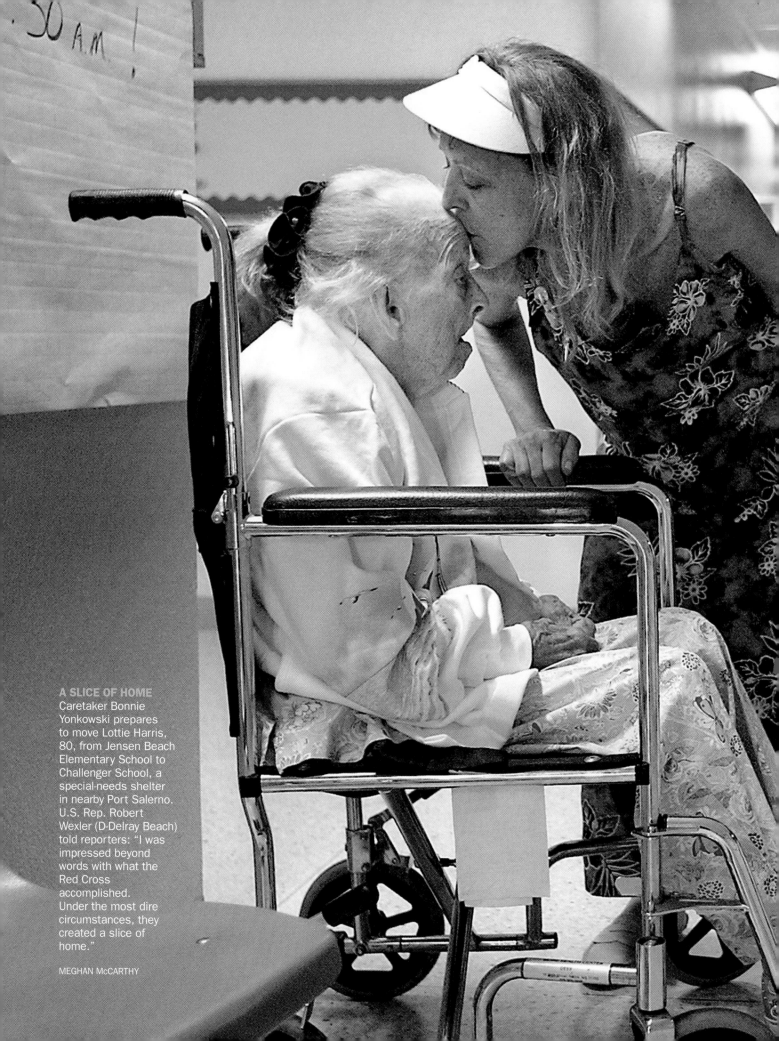

A SLICE OF HOME
Caretaker Bonnie Yonkowski prepares to move Lottie Harris, 80, from Jensen Beach Elementary School to Challenger School, a special-needs shelter in nearby Port Salerno. U.S. Rep. Robert Wexler (D-Delray Beach) told reporters: "I was impressed beyond words with what the Red Cross accomplished. Under the most dire circumstances, they created a slice of home."

MEGHAN McCARTHY

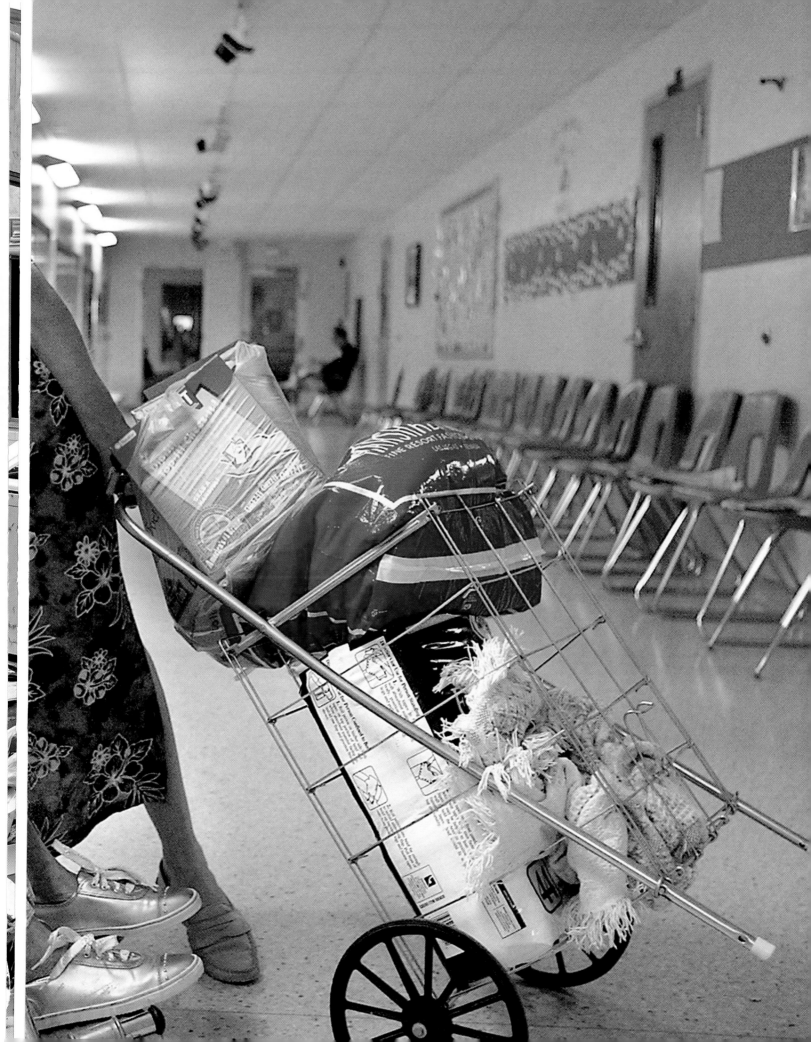

Previous pages: Frances destroyed the Fort Pierce City Marina, where fishing boats, sloops, catamarans and yachts such as the $3.2 million, 100-foot Lady Diane were stacked like kindling. One boat insurer said: "We have 104 boats here, and all but three of them are total losses."

GREG LOVETT

DAVID LANE

SPEECHLESS

Betty Munnell, 78, of Fort Pierce reacts to hearing that a friend's mobile home was destroyed by Hurricane Frances. "I don't like Florida," said one resident. "I never did, and I like it even less now."

EXPOSED

A living room set is visible inside a unit at Ambassador condominiums in Lake Worth after Frances blew out a wall. Hurricane Frances exposed weaknesses in buildings and in landscape.

STEVE MITCHELL

GREG LOVETT

OUT AFTER 48 YEARS

Vincent Sparaco, 86, rests as he cleans out his refrigerator after Frances. His Lake Worth home was badly damaged when a tree fell over onto the roof and allowed the rain to come in. He has lived in the house for 48 years, but the Palm Beach County Health Department said it is uninhabitable.

"This is Florida's 9/11"

STEVE BEDNER,
a grower with family-owned pepper and cucumber farm Bedner Growers west of Delray Beach

"We're the Alamo of citrus"

BOB ULEVICH
of the St. Johns Water Control District, which controls conditions for Indian River citrus

TAYLOR JONES

WORRIED FARMERS

It was standing room only (**above**) Sept. 8 for a meeting with state Agriculture Commissioner Charles Bronson. Florida's farmers and ranchers saw inundated fields, with some cattle standing in water up to their bellies. "That ($3 billion in losses) is a big chunk of money for agriculture to lose and for the state of Florida to lose," Bronson said of the industry that employs 300,000 and had more than $6 billion in revenues in 2003.

BITTER HARVEST

Previous pages: Hundreds of grapefruit float under flooded trees in the Sun Ag. Inc. groves in Indian River County on Sept. 6.

RICHARD GRAULICH

Agriculture cost: $3 billion

Florida's 2004 grapefruit harvest will be the smallest in 66 years: The Indian River region is expected to produce only 5 million 85-pound boxes of grapefruit, compared to 28 million last year. Statewide, the grapefruit harvest will be only 15 million boxes this year, the smallest since 1937-38. (Even at 15 million boxes, Florida produces more grapefruit than any other state.)

The orange crop, which reached a near-record 242 million 90-pound boxes in 2003-04, is expected to produce 176 million boxes this year.

More bad news: Citrus canker germs may have hitchhiked on Frances and Jeanne into much of central Florida.

Some good citrus news: The honeybell crop is expected to be higher than last season. The honeybells (a type of tangelo) were smaller on the trees than grapefruits when the storms hit, and fewer dropped to the ground.

– SUSAN SALISBURY

The big question: Who's got the power?

By

FRANK CERABINO

Palm Beach Post columnist

Originally published Sept. 8, 2004

"Do you have power yet?"

Nearly every conversation I've had during the past few days begins that way.

You may not agree that there are two Americas. But since Frances, everyone would certainly agree that there are two Southeast Floridas: the one that has electricity and the one that doesn't.

I am one of those people living in the dark, sweating inside a partially boarded up house, fantasizing that every big engine I hear is the arrival of an FPL truck.

But so far, the only rumbling engines have been generators from the few houses on the block that have spent the money and the time to insulate themselves from the discomfort caused by Hurricane Frances.

I don't begrudge these people. But I am liking them a little less each day, as their incessant loud engines have made it impossible for me to achieve even the minimal comfort of sweating in peace.

The worst part about this is that there are people with power all around us. It might be easier to bear if everyone were out of power. But power restoration has a maddeningly random quality to it, a kind of block-by-block voltage lottery that defies prediction.

"We'll have power some time tonight," one of our friends predicted Sunday night.

He and his family had joined ours for a candlelight dinner of spaghetti with jarred sauce, and Entenmann's chocolate-frosted doughnuts. I boiled the water for the spaghetti on my barbecue grill.

"In the middle of the night, it will just come on," he predicted.

But it hasn't, although I've imagined that moment many times.

Monday night, we ventured out for dinner into the world of light that is Broward County. It was a magical world where gas stations were pumping gas, traffic lights worked, and restaurants were open.

Getting dinner, though, was a challenge. As we cruised down Federal Highway, we found plenty of other families like ours. It wasn't until we were well into Pompano Beach before we found a restaurant that wasn't packed.

And by the time we crossed back into Palm Beach County, it was after the 8 p.m. curfew. A Boca Raton police officer blocked Dixie Highway, checking identifications, and letting only city residents pass.

"Go straight home," he told us.

We did, hoping once again to see the house lights blazing. No such luck.

We walked our dog through the darkened neighborhood, surprised to round the corner and discover lights on the next street, and hear the soft, sweet hum of central air-conditioning units.

"Don't we know anybody here?" I asked my wife.

We walked home. The house was too hot, so we sat on the front porch. My 4-year-old son brought his coloring book, and using a flashlight's beam, colored Ninja Turtles. I brought out a portable radio and tuned it to an evening jazz show that dedicated its post-hurricane show to nothing but the blues.

It was breezy outside, and there was something to be said for having nothing better to do than sitting with your family and listening to Ray Charles drown out the generators with his version of *Blues in the Night*, on a night lighted for us by only a couple of meager stars.

LANNIS WATERS

EXTENDING SOME HELP

An extension cord snakes across South 12th Street in the town of Hypoluxo, from Basil and Irene Zickafoose's home to Billie Roberts. Zickafoose said his neighbor asked him: "Why are you doing this? You don't even know us." He replied, "Because we can."

WEARY MOM
Jill Muolo sits in the screened carport of her Cabana Colony home in Palm Beach Gardens, the spot where her family spent most of their time since Hurricane Frances knocked out power. Ice and generators have been in short supply since the storm blew in Sept. 5. That's tough for everyone, but particularly hard when you have a baby who needs warm bottles. Muolo warmed baby bottles for her son, Christopher, over a fire.

GREG LOVETT

STRINGING AND RESTRINGING

City of Bartow Electric Department worker Gary McKinsey (**below**) repairs lines following Frances. The city, 30 miles east of Tampa, was hit by both hurricanes Charley and Frances . . . and then Jeanne. Almost 3 million Florida Power & Light customers were without power after Frances, and millions more who get power from other utilities were affected statewide.

GARY CORONADO

WELCOME SIGHT
Power crews (**right**) tackle the traffic lights at Parker Avenue and Belvedere Road in West Palm Beach on Sept. 8. FPL, which coordinated 1,400 workers, but sometimes left people in the dark about their progress, took heat. "They are doing their best," said state Sen. Ron Klein, D-Delray Beach. "But the communication side of this thing is a disaster."

GREG LOVETT

THE FINAL PUSH

Roberto Rubio and Ernesto Paredes (**above**), both of West Palm Beach, push Patricia Oliphant's car on Sept. 6 after she ran out of gas waiting in line at a Citgo station west of U.S. 441 on Okeechobee Boulevard. Gas supplies were low because tankers were waiting offshore until the hurricane passed.

GROCERY LINE

Customers at the Winn-Dixie (**left**) on 45th Street in Mangonia Park line up to get in the store on Monday morning, Sept. 6, the day after Frances. By 9:30 a.m., the store was out of ice and charcoal but still had four pallets of water.

A RUN ON ICE
A crowd jostles to get bags of ice at the Publix supermarket on Southern Boulevard in West Palm Beach. Each customer was allotted two bags of ice. It was gone in a half-hour.

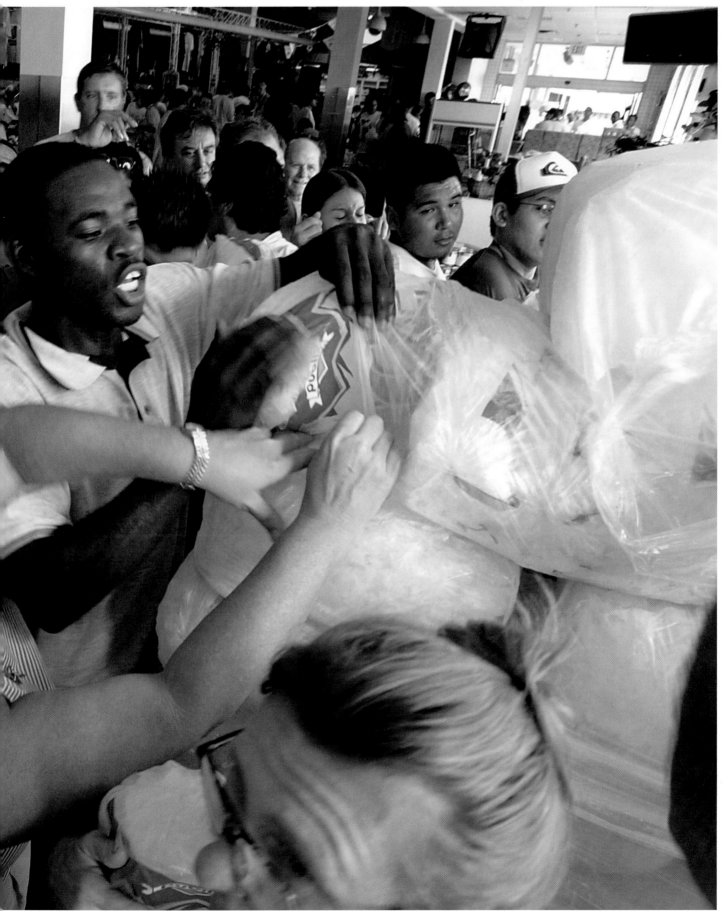

SUSANA RAAB

LONG WAIT
Makayla Oquendo, 4, of Greenacres, watches Sept. 6 as an Army convoy arrives at the South Florida Fairgrounds, one of Palm Beach County's hurricane relief distribution sites. Makayla was waiting with her mother for free ice, water and food. The Oquendos arrived at 8:45 a.m., but the supplies weren't handed out until 1 p.m.

ALLEN EYESTONE

JAY JANNER

GEE, THANKS, MR. PRESIDENT

President Bush hands a bag of bottled water to a victim of Hurricane Frances at a water distribution center at Lawnwood Recreation Area in Fort Pierce on Sept. 8. Bush made five visits to hurricane-damaged Florida in August and September. First lady Laura Bush and daughters Jenna and Barbara also visited, spending about a half-hour loading cases of water and ice with relief workers at the Indian River Mall in Vero Beach. The same day, Teresa Heinz Kerry, wife of presidential contender Sen. John Kerry, visited Riviera Beach.

FARM AID

The Guatemalan-Maya Center in Lake Worth became a staging area for immigrants seeking ice, water and baby formula. "They work in the fields," said Lucio Perez Reynozo, executive director of the center. "Not working two to three days is a crisis for them."

GREG LOVETT

"Once again, Florida has faced the devastation of a hurricane, and once again the people of Florida are showing their character."

PRESIDENT BUSH,
visiting the Treasure Coast on Sept. 8

VADA MOSSAVAT

TOURING DISASTER AREA
Gov. Jeb Bush arrived in West Palm Beach Sept. 5 to confer with emergency management officials and assure residents that "this will be a massive effort" to restore services. With him (from left) County Commissioner Karen Marcus, U.S. Rep. Clay Shaw, State Rep. Adam Hasner and U.S. Rep. Mark Foley. The governor won widespread praise for his leadership during Florida's hurricanes. U.S. Sen. Bob Graham called his performance "exemplary." A public opinion expert at Florida State University, Jay Rayburn, said: "If you look at the checklist of what ought to be done in a crisis situation, he'd get an A-plus." Before Charley made landfall, Bush asked his brother to declare the state a disaster area. President Bush complied. Speaking in English and Spanish, Gov. Bush cautioned, then reassured Floridians during the state's greatest calamity in decades.

Gov. Bush: Hurricane Andrew "scariest thing" he's ever lived through

Gov. Jeb Bush confronted four hurricanes in 2004, but none like the one he rode out in 1992.

As Hurricane Andrew threatened Miami-Dade County, the Secret Service asked Bush, then a local developer, and his family to leave their home, which was east of U.S. 1, and evacuate to Miami-Dade County's Emergency Operations Center. (Bush's father was president in 1992, so he had Secret Service protection.)

Bush refused to go. "I don't think it's appropriate for my family to be in some special place when there are 1.8 million people in our county not afforded that opportunity," he told them. The bodyguards had to leave because of the unsafe conditions.

"My dad called me and said, 'You sure you want to do this?' And I said, 'Yeah, we're fine.' "

Bush and his wife, children, mother-in-law and dog huddled in a hallway at a friend's home. "It literally felt like the house was going to explode out," he said. "The pressure was so high when the winds were the strongest, and it made a really weird sound. It was the scariest thing I've ever had to live through. There's no reason to want to live through a storm of that magnitude. It makes no sense at all."

If he knew then what he knows now, he'd evacuate, Bush said before Ivan hit this year. He advised Panhandle residents to do the same:

"Trust me, it is a powerful, powerful force of nature that you shouldn't be messing with."

– DARA KAM

THE SAGA OF "BETTY BOO"
After Skip Piker, 64, rode out Hurricane Frances in his sailboat Betty Boo in the Intracoastal Waterway near Riviera Beach's Phil Foster Park, rough seas tore his anchor lines and cast the vessel adrift. But waves drove the boat right onto the park's beach. When the seas finally calmed, Piker's boat was a landlubber, sitting keel deep in the sand, its bow a mere 15 feet from the forbidding concrete of the Blue Heron Bridge. After a friend towed the boat back to the water, an overjoyed Piker (**left**) prepares to swim out to it. Piker said he made it through the storm because "the devil don't want me, and God said stay the hell out." Piker began living on his sailboat four years ago, after his wife died of cancer.

THOMAS CORDY

UMA SANGHVI

SKY'S THE LIMIT
Roofers work on a house on Bilbao Street in Royal Palm Beach on Sept. 23, following Hurricane Frances but before Jeanne struck. The roofing crisis was bad for homeowners but good for contractors. "I went from two crews to six, one estimator to six," said Royce Arnold, whose roofing business suddenly went, well, through the roof. Some enterprising auctioneers even put their roof shingles up for sale on eBay. The Internet auction site featured "A Genuine Shingle Blown off My Roof by Hurricane Frances" and "Hurricane Frances Rain Water." The site read: "You are bidding on one vial of actual rain water from Hurricane Frances! Collect a piece of history from this one-of-a-kind weather event!"

FACE-TO-FACE WITH FEMA
Laurie Neill (**right**) of Palm Beach Gardens talks to Federal Emergency Management Agency worker Valente Benavides about the damage to her roof inflicted by Hurricane Frances. By late October, FEMA had doled out roughly $172 million in post-hurricane aid to residents of Palm Beach County and the Treasure Coast.

GREG LOVETT

"The best feeling in the world is to see that truck. When I see that truck, I light up like Christmas."

JANET BRIGHT
of Jensen Beach, talking about the American Red Cross truck

A friend in need

$92 million

What the Red Cross spent on Charley, Frances, Ivan and Jeanne relief, according to National Red Cross spokeswoman Laura Howe. The Greater Palm Beach Area Chapter alone spent at least $3.5 million.

The Red Cross
Greater Palm Beach Area Chapter
(serves Glades, Hendry, Okeechobee and Palm Beach counties)

Shelters operated during/after Frances: **33**
Shelters operated during/after Jeanne: **24**
Meals/snacks served: **807,562**
People sheltered: **35,760**
Mental health contacts: **2,279**
Disaster health services contacts: **2,528**
Family service cases opened: **2,450**
Volunteers: **919**
Staff: **56**

Martin County Chapter

Shelters operated after Frances: **8, housing 4,000 people**
Shelters operated after Jeanne: **5, housing 2,400 people**
Meals served: **307,931**
Mental health contacts: **over 2,279**
Family services cases opened: **1,242**
Volunteers: **1,000 (including 134 non-local)**
Staff: **12**

North Treasure Coast Chapter
(Serves St. Lucie and Indian River counties)

Shelters operated after Frances: **18, housing 9,000**
Shelters operated after Jeanne: **14, housing 4,000**
Meals served: **More than 500,000**
Mental health contacts: **4,200**
Family services cases opened: **4,755**
Volunteers: **451 (including 103 non-local)**
Staff: **4**

PASTOR PASSES THE BREAD
Rob Shrader, an associate minister at First Christian Church, brings donated dinners to residents of West Palm Beach's Christian Manor on Sept. 10. Five days after the storm, the senior community was still without power, and Shrader and his wife hauled dozens of boxed dinners up three flights of stairs.

HOT FOOD
The hurricanes inspired generosity. At the Florida Culinary Institute, about 60 chefs, culinary students and volunteers prepared more than 11,000 hot meals for Frances victims (here, they prepare pulled pork). Food-service company Cheney Brothers also supplied food for hurricane victims and worked with Continental Catering to prepare about 12,000 meals a day after Hurricane Jeanne.

UMA SANGHVI

JAY JANNER

Special delivery – in the dark

By

EMILY J. MINOR
Palm Beach Post columnist
Originally published Sept. 7, 2004

The baby came around 2:45 in the morning, Saturday going into Sunday, the hurricane raging. Finally.

It was a girl, 6 pounds, 11 ounces, and you know what they named her, right?

Frances.

"We couldn't send any units because the winds were in the 80 miles-per-hour range," said Palm Beach Fire-Rescue Capt. Don DeLucia. "They had to (deliver) in the dark. It was a real panic."

The real panic, of course, was on the other end of the phone line in suburban Lantana, when a 28-year-old mother of five called 911 shortly after 1:30 in the morning and got dispatcher Jill Anderson.

"She told me her water had just broke," said Anderson, who's been a dispatcher for 15 years and had previously delivered two babies by telephone. "She was actually pretty calm."

Anderson, 39, talked the woman's father through the delivery, step by step. The baby would be slippery, so get a towel. In fact, get several. Once the shoulders are out, the newborn will pop out fast. Grab an arm, wipe it off.

The mother didn't want their names published and has refused interviews. "The *Today* show called. So did the *New York Post*," said Madelyn Passarella, a spokeswoman at JFK Medical Center, where the woman and her healthy baby were taken when the winds calmed.

"She's cute as a button. I got to hold her."

THE CALM BETWEEN THE STORMS
Rose Smith, 30, of Royal Palm Beach unwinds from the stress of Hurricane Frances on Sept. 12 at Lake Worth Beach, the damaged pier stretching out behind her. But two weeks later, Hurricane Jeanne would come ashore on almost an identical track to Frances', inflicting more damage, more outages, more misery.

JAY JANNER

GREG LOVETT

An air conditioner in Bathtub Beach parking lot on Hutchinson Island.

DAVID SPENCER

The pool at Atlantis condominiums on Hutchinson Island.

LANNIS WATERS

A living room at Coquina Isle condo in the Panhandle, after Ivan.

LANNIS WATERS

A van engulfed by sand on Hutchinson Island.

GREG LOVETT

Residents of Hutchinson Island send up a plea after Frances.

ALLEN EYESTONE

A Hurricane-Jeanne-battered house on Satellite Beach.

Beachfront property, redefined

Welcome to Hutchinson Island, where first-floor condos come with an ocean view and a private beach . . . in the living room.

Island residents were the victims of a hurricane tag team: Frances washed away the dunes, allowing Jeanne to blow even more sand inside. Most of the island's beachfront condos had 4 to 5 feet of sand filling the first floor.

In the Panhandle, Ivan blew Pensacola's famed dunes away or up onto streets.

The sands on Florida's beaches are 60-70 million years old, made of powdered granite and silicates washed down from the Piedmont and Appalachian plateaus when dinosaurs still roamed the Earth. Our beaches are literal, glittering Jurassic Parks.

Now they have been blown inland, but that is no shock. Florida's beaches are dotted with sea walls, groins, jetties and towering condominiums. Against these concrete teeth, wind, water and tides break with full force, scooping out the sand behind them as they recede and return. Hurricanes speed up the process.

Jim Bounds (**below**) takes a break from digging sand out of his home near the Indian Hills Golf Course in Fort Pierce. The golf course was under construction, and when Hurricane Frances hit, many nearby homes were buried in sand meant for the sand traps.

JAY JANNER

Far right: A fire hydrant buried up to its neck shows the sand's depth in the parking lot at the Regency Island Dunes condominium on Hutchinson Island. Frances and Jeanne piled up to 5 feet of sand in the living rooms of Hutchinson Island's first-floor condos.

A family snapshot found in the Grande Lagoon subdivision of Pensacola, which was hit by a wall of water when Hurricane Ivan came ashore.

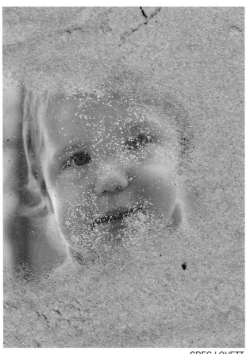

GREG LOVETT

LANNIS WATERS

GREG LOVETT

An aerial shot of Pensacola Beach on Florida's Panhandle shows how Hurricane Ivan buried streets and sidewalks near the shore with a solid coating of sand. Before Ivan, Pensacola was known for its sand. A message on a water tower at the edge of town brags: "Greater Pensacola: Florida's Finest Beaches."

RICHARD GRAULICH

DAVID SPENCER

It's easy to get into the crawlspace (**above**) now for maintenance man Mike Kilcrease (left) and resident Johnny Vars, who inspect the sand-filled pool house at the Sea Winds condominiums on Hutchinson Island following Hurricane Jeanne. Vars' father, Doc, rode out Hurricane Jeanne in his third-floor condo. "The bulldozers and backhoes had just come and pushed the sand back (after Frances)," Doc Vars said. "We just stood there and watched the waves eradicating all that."

Crabs (**left**) knock against the glass of the first-floor social area at Island Dunes condominium on Hutchinson Island. A foot and a half of sand was deposited next to the glass doors by Hurricane Frances.

WATER HAZARD
The Seminole Golf Club in North Palm Beach, submerged by flooding from Hurricane Jeanne, was just one of many golf courses dealing with fallen trees and washed-out bunkers. Hurricanes Frances and Jeanne cost South Florida golf courses millions in damages and lost greens fees. At hard-hit Loblolly Pine in Hobe Sound, director of golf Rick Whitfield said: "I just hope people understand this is a freak of nature."

ALLEN EYESTONE

"Thankfully, the integrity of the golf courses hasn't been lost, but some of the aesthetics won't be the same."

BUD TAYLOR,
director of golf at PGA Golf Club in Port St. Lucie,
one of many hard-hit area golf courses

TAYLOR JONES

GARY CORONADO

LIGHTS OUT
Hurricane Frances knocked over seven of the eight light towers around Roger Dean Stadium in Jupiter, spring-training home to the Florida Marlins, and caused about $2.7 million in damage. Hurricane Jeanne added to that tab when it blew off temporary roof coverings. Total damages will be at least $3 million from both storms, said Rob Rabenecker, the stadium's general manager.

PLAYIN' IN THE RAIN
Miami Dolphins defensive end Jason Taylor takes in not only the rain but the Dolphins' 13-3 loss to the Pittsburgh Steelers Sept. 26 at Pro Player Stadium. The game was delayed until evening after Hurricane Jeanne made landfall that morning along the Treasure Coast.

HURRICANE IVAN

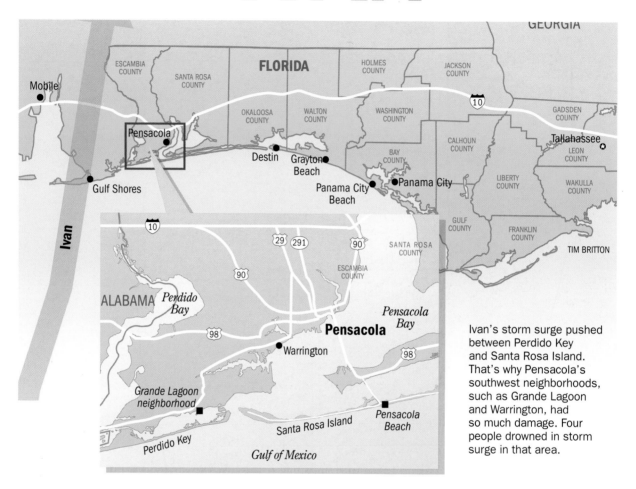

Ivan's storm surge pushed between Perdido Key and Santa Rosa Island. That's why Pensacola's southwest neighborhoods, such as Grande Lagoon and Warrington, had so much damage. Four people drowned in storm surge in that area.

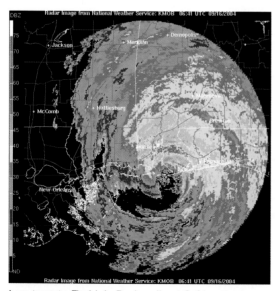

Ivan targets Florida's Panhandle at 2:41 a.m. on Sept. 16th. Most powerful winds hit Pensacola.

■ **Category 3 hurricane**

■ **Landfall:** Sept. 16, Gulf Shores, Ala. The strongest part hit Florida's Panhandle, causing tornadoes before the eye hit.

■ **Top sustained wind speed at landfall:** 130 mph

■ **Size of eye:** About 40 miles

■ **Notable characteristics:** A big storm with powerful storm surge (up to 16 feet) that spawned deadly tornadoes in Florida, Ivan was just 1 mph shy of being a Category 4. Ivan also caused widespread flooding in West Virginia and Pennsylvania. A remnant of the hurricane – a small swirl of low pressure – then broke off and slowly drifted south.

■ **Deaths:** 29 in Florida, more in Southeast U.S. and Caribbean

■ **Damage:** Up to $10 billion in insured losses

■ **Where deaths occurred**

Bay	2
Calhoun	4
Escambia	14
Manatee	1
Okaloosa	4
Santa Rosa	4

Source: Florida Division of Emergency Management, Oct. 19, 2004

POWERFUL WINDS
Annette Burton (**above**) visits Wayside Park in Pensacola on Sept. 15 to feel the power of Ivan. The Category 3 storm arrived early the next day, killing at least 29 people in Florida and demolishing waterfront homes.

BUCK STOPS HERE
Rusty Gilbert, Adrian Tannler and Brad Johnson (**left**) of the Alabama Gulf Coast Zoo in Gulf Shores, Ala., capture an Asian fallow buck that escaped from the zoo after Hurricane Ivan caused flooding up to 12 feet in places.

REDUCED TO RUBBLE
Previous pages: Debris is strewn everywhere in the Grande Lagoon subdivision of Pensacola.

GREG LOVETT

Right: Winds of 130 mph and storm surge of up to 16 feet flattened this waterfront home in Grande Lagoon.

GREG LOVETT

What killer storm surge can do

Four people drowned in Ivan's storm surge, including some residents of Grande Lagoon, an affluent waterfront neighborhood.

Mechelle Smith and her family survived – barely. When the water got to chin level, she jumped on the kitchen counter with her two young children and her parents. Then they sloshed out to the garage and climbed on top of their Plymouth Voyager van. When the van started to float away, the Smiths made a life-or-death decision: "We didn't know if we should make a run for it or go to the attic and higher ground. We didn't want to get stuck in the attic and drown." They chose the attic – where they stayed for four hours – and lived.

Of 400 homes in Grande Lagoon, 70 percent will probably have to be demolished.

After the storm, Mechelle Smith rummaged through her parents' wrecked house. "There was a bass in the drawer of my mother's curio cabinet," she said.

– JOHN PACENTI

COOL TRUCK
A truck (**right**) ended up in the swimming pool in back of a home in Pensacola's Grande Lagoon area, where Ivan's powerful winds and storm surge took their greatest toll.

GREG LOVETT

CHILDHOOD HOME DESTROYED
Alex Norton, 19, breaks down after seeing her childhood home (**right**) destroyed along Scenic Highway in Pensacola.

GREG LOVETT

GREG LOVETT

Death toll in Florida from the four hurricanes: 125

Roberto Alvarado's death got national attention because of photos of his truck hanging off the bridge. But most of the 125 direct and indirect deaths from the four 2004 storms were from accidents before or after the storms: generator-related carbon monoxide poisoning, falling off ladders or roofs, burns from fires started by candles, car accidents, contact with power lines.

At least 14 deaths were from heart attacks or medical conditions.

At least three of the deaths were suicides of people despondent over damage.

SYMBOL OF DESTRUCTION
Washington state trucker Roberto Molina Alvarado was killed Sept. 16, when the cab of his tractor-trailer plunged off an Interstate 10 bridge destroyed by Hurricane Ivan. This scene became a symbol of Ivan's destruction. Divers found Alvarado's body in Escambia Bay, where his cab landed. The bridge, near Pensacola, reopened Oct. 5, with two-way traffic sharing the westbound lane while repairs were made on the eastbound span.

What stood and what didn't

RICHARD GRAULICH

RICHARD GRAULICH

RICHARD GRAULICH

BUILT TO ANDREW STANDARDS
Townhouses (**top**) in the White Sands community on Pensacola Beach that survived Ivan were built to Hurricane Andrew standards, after older units were destroyed by Hurricane Opal in 1995.

BUILT IN THE 1970s
The townhouses (**middle**) built in the 1970s are a total loss after Ivan.

TWO HOUSES, TWO FATES
Two houses (**left**) on Belmont Street in the Old East Hill Historic District of Pensacola had distinctly different fates when Ivan hit. The house on the right was built in 1903 and refurbished. The house on the left was built by Habitat for Humanity and is only a few years old.

"A BLESSING TO BE ALIVE"
Lillie Mae Bogan, 82, sits on her front porch **(below)** in front of her Pensacola home, which was leveled by Ivan. She tried to ride out the storm as the house blew apart around her. She finally took refuge in her car. As her relatives sorted through the home's debris, Bogan asked them to find a Bible. "It's a blessing to be alive," she said. "I only have one life."

GREG LOVETT

RICHARD GRAULICH

LOOK, MY MEDAL!
Patrick Thrasher, 12, finds his Space Camp medal on Sept. 21 while helping his folks sift through the rubble of their former home on Seaglades Drive in Pensacola.

GREG LOVETT

"CATASTROPHIC"
A bicyclist **(left)** wends his way along a road near Pensacola's waterfront, which the city had spent millions redeveloping. Ivan chewed up asphalt, unearthing buried electric cable. "It's catastrophic. The electric system it has taken us 80 years to build was basically destroyed in eight hours," spokesman John Hutchinson said.

LOOKS LIKE A SANDBOX, BUT IT'S A ROAD
Residents of Pensacola Beach walk along Fort Pickens Road Sept. 22. Some people walked for miles along sand-piled roads before seeing the damage to their homes. The finely powdered sands of the beaches here once rose in beautiful dunes, but those dunes are gone now, washed away (as was most of Pensacola's recent $16 million beach reclamation project). Storm surge hit hard in suburbs southwest of Pensacola. In Warrington, homes appeared to be lifted from their foundations and set back down. The surge scenario played out repeatedly: ground floor decimated, second floor untouched. One resident said the water level went up 13 feet at his home a block from the bay. "Pensacola may never be the same," City Manager Tom Bonfield said.

GRIEVING BROTHER

J.W. Marshall of Blountstown (**right**) sobs as he touches the caskets of his brother, James Harold "Bull" Marshall, 41, and sister-in-law Mary Lee Marshall, 37, who died when a tornado flung their mobile home off its foundation.

GREG LOVETT

SAD FAREWELL

More than 500 people crowded into the First Baptist Church in Blountstown, near Tallahassee, for the funeral of James Harold "Bull" Marshall and Mary Marshall (**above**). The couple, who worked for the state Department of Corrections, died when a tornado struck their mobile home, tossing it across State Road 69A. (The same twister killed James Terry and Donna Terry-Reed, a father and daughter who lived nearby.) Officers from the Calhoun Correctional Institution were given the day off to attend the funeral. They wore black bands across their badges. "If you needed something taken care of, you called the Bull. He was very much a man," said Sgt. Andy Williams, who worked with the Marshalls and described them as family. "There's no color or gender here today. We're all brown," he said, referring to the officers' uniforms.

GREG LOVETT

HURRICANE VETERAN
Diane Kelly of West Palm Beach stands in line at the Home Depot on Palm Beach Lakes Boulevard on Sept. 23. She's stocking up on 5-gallon gas containers to power her generator in case Hurricane Jeanne knocks out power the way Frances did just three weeks before. The lights did go out, leaving 1.7 million Florida Power & Light customers in the dark. But electricity was restored faster after Jeanne than Frances. Half of Palm Beach County's customers had their power restored just two days after Jeanne hit. The Treasure Coast was in the dark longer because Jeanne knocked down some concrete utility poles and snapped some wooden ones.

GREG LOVETT

HURRICANE JEANNE

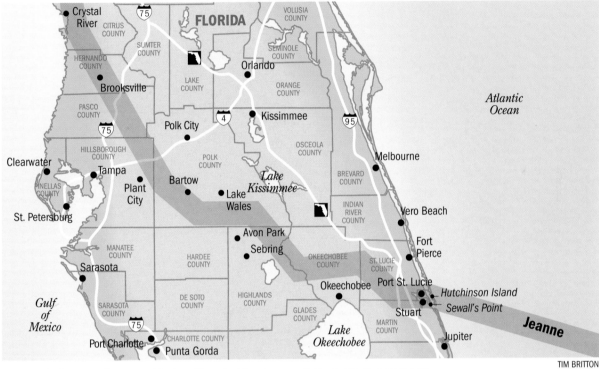

Crystal River
CITRUS COUNTY
FLORIDA
VOLUSIA COUNTY
SUMTER COUNTY
HERNANDO COUNTY
Brooksville
SEMINOLE COUNTY
Orlando
LAKE COUNTY
ORANGE COUNTY
PASCO COUNTY
Polk City
Kissimmee
Clearwater
HILLSBOROUGH COUNTY
Tampa
PINELLAS COUNTY
Plant City
Bartow
Lake Wales
Lake Kissimmee
OSCEOLA COUNTY
BREVARD COUNTY
Melbourne
St. Petersburg
INDIAN RIVER COUNTY
Vero Beach
Avon Park
Sebring
MANATEE COUNTY
HARDEE COUNTY
OKEECHOBEE COUNTY
ST. LUCIE COUNTY
Fort Pierce
Sarasota
Okeechobee
Port St. Lucie
Hutchinson Island
Sewall's Point
Stuart
Gulf of Mexico
SARASOTA COUNTY
DE SOTO COUNTY
HIGHLANDS COUNTY
GLADES COUNTY
MARTIN COUNTY
Jupiter
Jeanne
Port Charlotte
CHARLOTTE COUNTY
Punta Gorda
Lake Okeechobee
Atlantic Ocean

TIM BRITTON

Jeanne hit Treasure Coast residents with a double-whammy, following in Frances' path.

■ Category 3 hurricane

■ Landfall:
Sept. 25, south end of Hutchinson Island

■ Top sustained wind speed at landfall: 120 mph

■ Size of eye: 35 to 40 miles

■ Notable characteristics:
Track made a loop-the-loop in the Atlantic before churning due west toward the Bahamas and Florida. It moved almost twice as fast as Frances and followed almost the same path across land.

■ Deaths: 16 in Florida. At least 2,000 in Haiti.

■ Damage:
$6 billion in insured losses

■ Where deaths occurred

Brevard	1
Clay	1
Hardee	1
Indian River	1
Lake	1
Miami-Dade	1
Orange	3
Palm Beach	2
Pasco	1
Pinellas	1
Polk	2
St. Lucie	1

Source: Florida Division of Emergency Management, Oct. 19, 2004

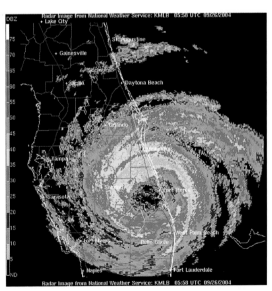

Jeanne's radar image at 1:58 a.m. on Sept. 26. Landfall came just before midnight on Sept. 25. Heavy bands of wind hit Palm Beach County, but areas on the north side of the eye wall got 10 hours of punishing winds with no respite from the eye.

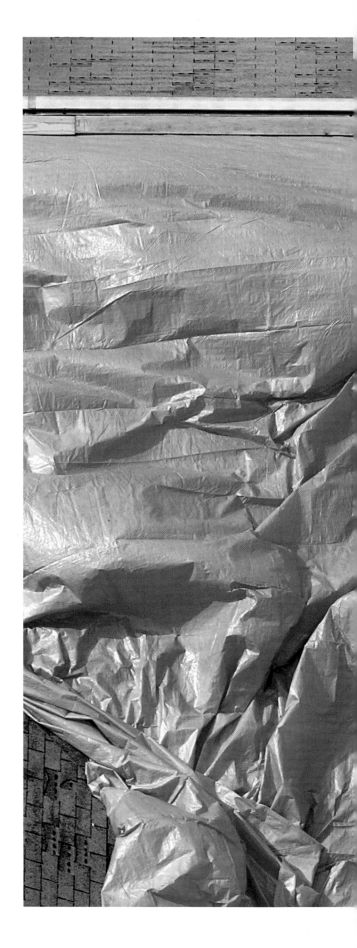

ANOTHER STORM BREWING
Workers in Hobe Sound (**right**) battle gusts from an approaching Hurricane Jeanne as they secure a tarp on a roof damaged by Frances. The U.S. Army Corps of Engineers' Operation Blue Roof covered at least 3,000 roofs in Palm Beach County with the blue plastic sheeting.

GREG LOVETT

HURRY-UP CLEANUP
Hugo Morales (**above**) works Sept. 24 to scoop up debris left over from Hurricane Frances in the Flamingo Park neighborhood of West Palm Beach. Crews across the area were rushing to clear streets before the impending strike of Hurricane Jeanne.

For all four storms, Florida's emergency response team delivered:

12,500
shipments of essential commodities (hauled in more than 4,400 trucks)

78 million pounds of ice

9.75 million gallons of water

13.8 million Meals-Ready-to-Eat packages

550,000 tarps

$1.7 billion
The federal/state price tag for these deliveries as of October

Source: Florida Division of Emergency Management State Emergency Response Team

GREG LOVETT

HERE COMES THE BRIDE . . . AND JEANNE
Russhelle Curry of Clewiston touches up her makeup at the Clewiston Inn as a TV set in the background shows Hurricane Jeanne approaching. Friends and family rushed to the inn to celebrate her wedding before the hurricane arrived late that night.

BILL INGRAM

THE HURRICANE CHANNEL
Residents of the Morse Geriatric
Center in West Palm Beach watch
nonstop coverage of Hurricane
Jeanne on Sept. 25 at about 6 p.m.
At 11:30 p.m., the power went out,
but the center provided food and
shelter for staffers and 82 family
members.

RICHARD GRAULICH

One home, two hurricanes

TAYLOR JONES

AFTER FRANCES . . .
This home on Orchid Island near Vero Beach lost its porch and siding during Hurricane Frances, and the storm surge ate away at the foundation. Though it looked OK from the outside, damage from Frances made this home unsafe.

. . . AFTER JEANNE
Hurricane Jeanne finishes off what Frances started, delivering the one-two punch that knocks out the home (**right**). Orchid Island was on the northern edge of both hurricanes, which meant it experienced the strongest, most unrelenting winds for approximately 10 hours. Gusts were up to 120 mph.

BOARDS WALK
Pilings from a destroyed boardwalk at Avalon State Park on North Hutchinson Island – Hurricane Jeanne's doing – seem to march across the sand.

DAMON HIGGINS

DAVID SPENCER

IT'S A WRAP
The tail section of a Sabreliner corporate jet is wrapped with sheet metal from the Galaxy Aviation hangar at the Witham Field airport in Stuart.

DOCUMENTING THE DAMAGE
Lourdes Cuadra (left) and daughter Lily Cuadra, both of Miami, document the effects of Jeanne at the Island Crest building on Hutchinson Island.

An awning blown off a mobile home at the Spanish Lakes I community in Port St. Lucie rests in a nearby palm tree.

SHANNON O'BRIEN

MEGHAN McCARTHY

First-floor units were filled with sand. "Oh, those poor people on the ground floor," said Lourdes, who owns a unit on the eighth floor. *Ay Dios.*

DAVID SPENCER

ALL FALL DOWN
A collapsed concrete-block wall, sheared off by Jeanne's powerful winds, crushed a van parked for safety behind the Stuart Shopping Plaza.

WADING DOWN THE HALL

As rain and flood water from Hurricane Jeanne began to fill their low-lying Palm Beach Gardens home, Jane Shea (**right**) tried to save her and husband, David's, belongings. "I was frantically pulling my husband's shoes and tossing them" into the closet, she said, "but the water was coming in underneath the foundation." The entire property ended up under 3 feet of water.

THE ASPHALT CRUMBLES

Surrendering to the pounding it took from Jeanne, the old Jensen Beach Causeway (**below**) collapsed like an overbaked cake. Department of Transportation workers (right) survey the damage where large portions of the road got washed away. Above is the new causeway.

THOMAS CORDY

MEGHAN McCARTHY

DAVID SPENCER

WAITING FOR A CREW
William Malphus, who owns BSB Trucking, takes a break from cleaning up the mess Hurricane Jeanne made of the Heritage Inn restaurant and lounge in Hobe Sound. Malphus of Springfield, Ga., said most of his crew left before Jeanne made landfall. He was doing minor repairs while he waited for them to come back.

COWS CROSSING
Cattle (**above**) look for higher ground after Hurricane Jeanne flooded the 1,200-acre ranch of Iris Wall in Indiantown. Out of 700 head of cattle, only one died in the storm.

KNIGHT IN SHINING TRACTOR
At Springhaven Estates near Indiantown, the only road in and out is impassible, even for heavy-duty pickups. Luckily for Janette Brewer, De Garrett came long in his tractor (**right**) and pulled her out to dry pavement. Brewer was bringing home Meals-Ready-to-Eat and other relief supplies.

WATERFRONT PROPERTY
A Jeanne-made lake at the Cypress Bay Mobile Home Park in Fort Pierce forces people to use a canoe to get to their homes. One resident's comment: "We're living. That's about it."

RIVER RUNS THROUGH IT
Meg and Bud DeFore of Fort Pierce wade through waters from the north fork of the St. Lucie River that flooded their home in the Rain Tree Forest neighborhood.

THOMAS CORDY

NO PARTYING

The cost of partying went up after Hurricane Jeanne. Countywide curfews were in effect from 10 p.m. to 6 a.m., and Palm Beach County sheriff's deputies arrested 132 people on Sept. 26. **Above**, Sheriff's Detective Mike Bianchi detains a curfew offender at Forest Hill Boulevard and Military Trail. When Sheriff Ed Bieluch announced his plan to lock up anyone who violated his curfew, he warned night crawlers: "Bring your toothbrush and your teddy bear."

GREG LOVETT

NO POWER

As the sun sinks in the west, the intersection at Southern Boulevard and South Olive Avenue dips into darkness in West Palm Beach. The power outage caused by the second hurricane to arrive in three weeks was by now a familiar scenario to area residents, who had learned to treat darkened signals as a four-way stop.

MUCH MULCH
Adam Carr, a bucket-truck operator from St. Petersburg, uses a chainsaw to cut away branches from the stump of a 100-year-old banyan tree at Boca Raton City Hall. In all, 50 football fields' worth of mulch were created by the four hurricanes in Florida.

BOB SHANLEY

ATTENTION, CORVETTE LOVERS
If you live on a barrier island, don't leave the car parked at the condo during a hurricane. This Corvette is outside the Ocean Rise condominiums on Hutchinson Island.

RICHARD GRAULICH

DAVID SPENCER

DAMON HIGGINS

WOODED LOT
This two-by-four plank was apparently driven by Hurricane Jeanne's 100-mph winds into the blacktop of the parking lot at Avalon State Park on North Hutchinson Island.

SCARY NIGHT
Karen Crow (**left**) reacts to her ruined Cinnamon Tree apartment in Jensen Beach. The roof, attic and ceiling collapsed during Hurricane Jeanne, trapping Crow and daughter Kelsey in the laundry room. Crow placed a frantic cellphone call to her brother. "I heard the terror in her voice," said Dennis Crow, who drove from Stuart and had to kick in the door to get to the two and their frightened dog.

BAD EGGS?
Winn-Dixie employee Richard Williams (**right**) throws away eggs at the grocery store on 45th Street in West Palm Beach. The store was without electricity for only one day after Hurricane Jeanne, but the chain wasn't taking a chance.

FAIR CATCH
Gabrielle Fitzgerald, 8, (**below**) helps hand out Meals-Ready-to-Eat at the South Florida Fairgrounds near West Palm Beach on Sept. 28. The Federal Emergency Management Agency shipped 1.6 million gallons of bottled water, 8 million pounds of ice, 1.1 million ready-to-eat meals, 770,000 containers of baby formula and 30,000 tarps to the areas hardest hit by Frances and Jeanne.

GARY CORONADO

BILL INGRAM

GREG LOVETT

"WE'RE VERY ANXIOUS"
Residents of Orchid Island (**above**) endure a frustrating wait before being allowed across the New Merrill Barber Bridge to check the damage to their barrier island homes, off Vero Beach. "We're very anxious," said Paul Mitzi, who had driven all the way from Illinois to check on the beachfront property of his 92-year-old aunt.

RELIEF FOR RELIEF WORKER
Diane Cuomo (**left**), a volunteer at the Port St. Lucie City Center, where ice, tarps and water were being distributed to victims of Hurricane Jeanne, rests a bag of ice on her head to cool off.

BLUE ROOFS
Next pages: This neighborhood near Donald Ross Road and Interstate 95, photographed on Oct. 13, bears the scars of two hurricanes.

GREG LOVETT

LANNIS WATERS

LIBBY VOLGYES

LOVING ARMS
(**above**) Austin Mathison, 7, (left) and T.J. DeLuca, 10, help comfort newborn Tera DeLuca as she cries in the arms of her mother, Teri DeLuca of Hobe Sound. "She's a little warm," said DeLuca.

RENT STRIKE AND A SIT-IN
Six-month-old Jonkerya Judd (**right**) sits in the lap of her mother, Lorraine Williams, outside their apartment on 12th Street in West Palm Beach, where the ceiling caved in after Hurricane Jeanne.

UMA SANGHVI

GARY CORONADO

$150 PER LOAD OF SCRAP METAL

Where do mobile homes and patio enclosures go after hurricanes get through with them? To the scrap heap. At Palm Beach Metal Inc. in suburban West Palm Beach, sellers of scrap metal have been cashing in, collecting around $150 per load. By Sept. 28, piles and piles of scrap filled the place.

TIARA'S SPARKLE GONE

The 42-story Tiara (**left**) rises like an apocalyptic vision on Riviera Beach's Singer Island. The once-elegant high-rise, built in 1977, now stands with its ribbing exposed. Hurricanes Frances and Jeanne sheared off whole sections of its synthetic stucco skin. The building predated building codes stiffened after Hurricane Andrew altered the landscape of Dade County in 1992.

DAMON HIGGINS

The amazing adventure of Priscilla White and her eight cats

By
CHRISTINE EVANS
Palm Beach Post staff

SUSANA RAAB

SHE FLED SO FUR WOULDN'T FLY
Priscilla White returns to the Ocean Breeze Trailer Park after riding out Hurricane Jeanne in a motel in Fort Lauderdale with a canary and eight Persian cats.

Theory: After four hurricanes in six weeks, Floridians are stressed out, worn out, roofless, rootless and a dang bit crazy. But they are also curiously resilient.

Proof: Priscilla White. Her story features a black Monte Carlo, eight Persian cats, a yellow canary, a car wreck, a Super 8 Motel, an Andrew Wyeth painting, a cat escape, a cat bite, a ruined oceanfront trailer, a break-in and a black eye.

It begins with her Sept. 25 evacuation of her home near Jensen Beach. White, 58, fled the Ocean Breeze mobile home park — her trailer is really just a way station for her cats, who must be flown to her second home on the Dutch Antilles island of Bonaire — and headed south to Fort Lauderdale in her black Monte Carlo, eight Persian cats and one yellow canary by her side.

"I went down the highway just fine," she reports. "Then I turned off and started heading to the Super 8. I was almost there when I got rear-ended. The Monte Carlo was totaled, and I was so upset. Now I've got this swollen face and black eye. They wanted me to go to the hospital, but it's 2:30 p.m., the wind is howling, and I didn't want to abandon my animals."

A tow truck takes the demolished car to the Super 8 so White can unload its contents, which she says includes an original Wyeth from her personal collection. The glass on the Wyeth is shattered from the crash.

She takes it into the motel room and places it atop the TV. She sneaks all eight cats in. She rides out the storm.

The next day, she rents a tiny white car for the "extortion" sum of $2,000, piles in the cats and the canary — but forgets the Wyeth.

She goes home to her trailer in the Ocean Breeze park, which is ripped like old curtains by Frances and Jeanne.

When she arrives, she opens her car door and one of the cats hops out. A good neighbor chases after it, nabs it — and suffers a wicked cat bite. The furry Persian hangs onto his hand like a coat hanger. The lacerations are deep, and an emergency room trip is planned.

But first, White drives down the street to her trailer. The latest hurricane has peeled off the aluminum roof like a sardine-can top.

The front porch is gone.

Inside, she spots a red brick. Now, she's feeling it, this dreaded sense of victimization. She has been robbed.

In her bedroom, on her dresser, four open jewelry boxes are stripped bare.

White tallies the damage:

She is missing one Andrew Wyeth, four boxes of jewelry, an aluminum roof, a front porch, at least one cat.

The one who bit her neighbor is hiding down the street. And, White fears, she might have left another at the Super 8.

She could cry but doesn't.

Because, like so many in this rain-soaked state, she is a survivor.

It could be worse. "I could be in Haiti."

Like so many others, Priscilla White plans to persevere.

"Nobody is saying, 'I'm not coming there, I'm going to Cleveland.'"

MAC McLAUGHLIN,
head of the Palm Beach County convention and visitors bureau

Back to the beach

Orlando got the winds of three hurricanes in 2004, but visitors to Disney World barely noticed. Tourist spots in the Panhandle took a bigger hit, but by the winter season, all of Florida was looking pretty much like paradise again.

At the 279-room Palm Beach Gardens Marriott, General Manager Roger Amidon wasn't concerned.

"I think we're all going to rebound very quickly," he said. "When the snows starts to fall up north like it did last year, people will be calling."

Added Charles Lehmann, director of the Palm Beach County Tourist Development Council: "Other than a few blue tarps on roofs, it doesn't look that bad."
— PAUL OWERS

SHANNON O'BRIEN

NO LINES TODAY
Spaceship Earth is shrouded in rain as Hurricane Jeanne blows through EPCOT theme park. Walt Disney World sent guests to their rooms, but theme parks opened the next day.

JUST A DAY AT THE BEACH
Carefree-looking beachgoers in Indialantic contrast with the debris. Between the two of them, Frances and Jeanne ripped about 6 million cubic yards of sand off Brevard County's shores: enough to fill the Cocoa water tower more than 8,640 times.

LANNIS WATERS

LANNIS WATERS

UH-OH, SPAGHETTI-OS FOR BREAKFAST

Alex Rediker, 5, enjoys a breakfast of Spaghetti-Os (**above**) while his dad, Steve, cleans up outside their home at the Ocean View apartments on Hutchinson Island. Steve Rediker said the apartment was flooded up to 6 inches deep by storm surge from Jeanne several days earlier.

FRONT-YARD SWIMMING HOLE

Kevin Tvenstrup, 13, does a flip in a rain-filled drainage ditch in front of his home (**right**) in The Acreage in suburban West Palm Beach. Four hurricanes severely tested the Water Management District's 511 miles of canals that control the level of already rain-swollen Lake Okeechobee.

ALLEN EYESTONE

CHRIS MATULA

WELCOME BACK!
History teacher Lu Haag gives Boca Raton Middle School custodian Luis Santana a welcome-back kiss as classes finally resume a week after Hurricane Jeanne knocked out power to the county.

Post-hurricane pop quiz

When did things get back to normal?

A. When the malls reopened.
B. When folks could get back on the Internet.
C. When kids went back to school.
D. All of the above.

Answer: D, of course.

So, it's back to the books . . . and homecoming, football games and fun. And even the dreaded Florida Comprehensive Assessment Test. Gov. Jeb Bush said county school administrators could delay the FCAT but not dump it this year.

Routines returned. Resiliency triumphed.

As St. Lucie Schools Superintendent Michael Lannon said when his schools finally reopened in early October: "We're down, but we're getting up."

Palm Beach County:
Missed 12 days of school

Martin County:
Missed 14 days of school

St. Lucie County:
Missed 20 days of school

Here's looking at you, storms

GARY CORONADO

Gulfport, Miss.: Homeowners mock Ivan with a painting on the front of their house.

DAVID LANE

Stuart: Don't mess with this guy.

White City: This homeowner makes it clear he has nothing to lose.

CHRIS MATULA

BRUCE R. BENNETT

West Palm Beach: Counting the days until power returns.

GREG LOVETT

Lantana: A sign on the Premiere Aviation hangar.

MEGHAN McCARTHY

Stuart: A challenge to Frances, hanging at Stuart Gardens Apartments.

ALLEN EYESTONE

Vero Beach: Ms. Pac-Man marks the path of Jeanne.

DAVID SPENCER

Stuart: National Guard vehicle is a veteran of all four storms.

JEAN SHIFRIN

Jensen Beach: Kate and Jeremy Harper try to lure FPL workers.

DAVID SPENCER

Sewall's Point: Matt McGrath's fish used to be mounted.

Bull's-eye:
The storms from space

Why did so many hurricanes hit Florida in 2004?
One factor was a persistent high-pressure ridge
in the western Atlantic,
which tended to turn the storms to the west.

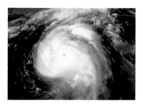

CHARLEY
The hurricane is along the west coast of Florida in this satellite image taken at 1:54 p.m. on Aug. 13. Charley was the most powerful storm of 2004 and the smallest in size.

NOAA

IVAN
Florida's Panhandle is the land mass at the bottom of the photo, taken Sept. 15. The Florida peninsula is to the left and Cuba and Mexico are the land masses at the top of the photo. Ivan strengthened to a Category 5 three times but weakened to a strong Category 3 (1 mph away from a Category 4) at landfall.

FRANCES
Frances is shown over Florida on Sept. 5. It was so big its tropical-storm-force winds hit Orlando International Airport and Miami International Airport at the same time.

JEANNE
This Sept. 25 satellite image shows Jeanne moving over Florida's east coast. Like Frances, Jeanne was a big storm, but it was more powerful.

FLORIDA'S HURRICANE HISTORY

Nature makes its own odds: Any season can be Mean Season

By
ELIOT KLEINBERG
Palm Beach Post staff

It's easy to forget about nature's power until it becomes your problem.

If it doesn't hit you, it doesn't happen? Right? That's folly, of course, but residents of Palm Beach County and the Treasure Coast were able to fool themselves for an incredibly lucky 25 years.

In 2004, we stared head-on at nature's fury, and nature knocked us to the mat.

This state is in a decades-long era of more and stronger storms. The researchers who concluded this said so three years ago, in July 2001. People just weren't listening then.

In fact, we're already nearly halfway through that active-hurricane era.

From 1995 to 2003 there were:

122 named storms (a record)

69 hurricanes

32 major hurricanes (Category 3 or higher)

But only three of those 32 major hurricanes hit the U.S., and not one hit Florida's peninsula.

We got spoiled. Now, we've received an incredibly expensive wake-up call.

From 1949 to 2003, no major storms hit Palm Beach, Martin, St. Lucie or Okeechobee counties. In that time, the population increased 10 times. Property values skyrocketed — jumping tenfold just in the past 25 years. That's why 2004's storms were so costly: More people plus higher property value equals more damage.

Floridians can never relax, not even in periods of below-average hurricane activity. Below-average years produced America's costliest hurricane (Andrew, 1992), America's deadliest hurricane (the Galveston hurricane of 1900) and America's most powerful hurricane (the Great Labor Day Storm of 1935).

Mean Season 2004 proved that statistics and averages mean nothing to the hurricane gods.

Ultimately, the odds of a hurricane hitting us remain 50-50.

Get used to it.

THOMAS CORDY

STORMY 2004
A palm tree bends in the outer bands of Frances as the hurricane heads into Florida on Sept. 4. The tree is framed in a broken window of a public restroom at Currie Park in West Palm Beach.

For the record books

2004: The first time Florida has been smacked by four hurricanes. First time for any state in more than a century.

1964: Three hurricanes hit Florida. Cleo (Aug. 27-29) tore up the east coast, Dora (Sept. 9-11) hit Jacksonville, and Isbell (Oct. 14) crossed the peninsula in a northeast track, pounding Palm Beach County.

1926: Three hurricanes hit Florida. The great Miami hurricane (Sept. 18) was preceded by a Daytona Beach storm (July 27) and a late-season storm (Oct. 20) that paralleled the South Florida coast and brought hurricane force winds ashore.

1886: The last time a state (Texas) was hit by four hurricanes.

Florida's most horrific storms

The storms of 2004 do not compare to the worst of nature's fury.

Hurricane of 1928

Florida's deadliest hurricane, and the third deadliest natural disaster in U.S. history

Sept. 16, 1928

Category 4

It was officially unnamed when it hit, but now it is legend: The Great Hurricane of 1928 killed at least 2,000 when the waters of Lake Okeechobee spilled out and flooded lake towns. The storm changed the way Lake Okeechobee and the Everglades were managed. The U.S. Army Corps of Engineers constructed a 150-mile dike around the lake to protect people from future hurricanes, though none, yet, has been anywhere near as powerful as the 1928 hurricane.

For an eyewitness account of Florida's deadliest hurricane, see page 136.

Labor Day Storm of 1935

Most powerful storm ever to hit the United States

Sept. 2, 1935

Category 5

Experts estimate winds of 150 to 200 mph hit the Florida Keys, with gusts exceeding 200 mph and a storm surge of 20 feet. The winds destroyed the Florida East Coast Railroad tracks to the Keys and killed at least 409 people, including hundreds of World War I veterans who were building bridges as part of President Roosevelt's New Deal program.

Hurricane Andrew

Costliest hurricane in history

Aug. 24, 1992

Category 5

Andrew smacked Homestead with top sustained winds of 165 mph. Andrew killed 26 people directly (another 62 indirectly), and caused $15.5 billion in damage in 1992 dollars (more than $40 billion adjusted for 2004 inflation). The Insurance Information Institute says Hurricane Andrew is the second costliest catastrophe in U.S. history, after the Sept. 11 terrorist attacks.

How do the 2004 storms compare?

Charley, Frances, Ivan and Jeanne combined are expected to cost insurers around $23 billion, according to the New York-based Insurance Information Institute. That's approximately half of their total cost, since insured damages are usually around half of the actual damages. More than one of every five Florida homes got some hurricane damage, and more than 2 million insurance claims are expected to be filed. (After Andrew, 700,000 claims were filed.) Florida Chief Financial Officer Tom Gallagher, who oversees insurance regulation, said: "These four storms are a much larger problem than Andrew was for the state."

MIAMI NEWS FILE PHOTO

Train swept off track by Labor Day Storm of 1935.

Other notable storms

Deadliest U.S. disaster: Galveston, Texas, hurricane of 1900, a Category 4 that killed 6,000 to 10,000.

Last hurricane to kill 100-plus in the U.S.: Agnes, a Category 1 that hit northwest Florida and the northeast U.S. in 1972.

Most U.S. deaths since Agnes: Floyd (1999), a Category 2 that hit the Carolinas and New England and killed 69 people.

Last hurricane to strike Palm Beach County and the Treasure Coast before 2004: David, a Category 1, in 1979.

Last major hurricane to strike Palm Beach County and the Treasure Coast before 2004: A Category 3 storm that killed 50 in 1949.

Only hurricane to strike North Florida in the 20th century: Dora, a Category 2 that struck Jacksonville in 1964.

Last major storm to hit Tampa Bay: An unnamed Category 4 that hit in 1921.

Top winds of more than 200 mph: Camille (1969), which hit Mississippi. Camille, Andrew and the 1935 Labor Day storm are the only Category 5 storms to strike the U.S.

Q&A with Max Mayfield

Max Mayfield is an unlikely media star, but when he talks, people listen.
Named director of the National Hurricane Center in 2000,
Mayfield had not overseen a major hurricane (Category 3 or higher)
striking Florida until this year.

ALLEN EYESTONE

Max Mayfield surveys Hurricane Jeanne damage.

What was the strangest thing about the season of 2004?

We had four hurricanes strike. Four is about as strange as it gets. And they were not weak hurricanes, either . . . I'm not surprised that Florida got struck. It's just that we had four.

Were you surprised by what you saw from the air?

The one thing that really did surprise me is that we did not have the tremendous storm surge in Charley. That should have been a tremendous storm-surge event. There are a lot of people who should be very, very thankful here that they did not get the storm surge.

Were you surprised by Floridians' behavior during the storms?

Overall, the nation's hurricane program came through with flying colors. The tremendous team effort between the meteorologists, the Hurricane Center, emergency managers, local officials, the media. In Florida, in particular, we should really be proud of the way we communicate with each other. We were all on the same page. People like to be told what to do; when to put their shutters up, when to evacuate. The recovery never goes as fast as people want, but in my opinion, we're light years ahead of Andrew.

What's the one thing you want everyone to know about hurricanes?

Here in Florida, people who experience hurricanes will now know that the people who had a hurricane plan did better than the people who did not. The people who had to figure something out at the last minute, they were stressed out for the first hurricane, but after the second, it was too much. I hope we've learned the lesson.

Do people recognize you on the streets now?

Yes, a lot more so than before. I guess I did get more visibility.

– ELIOT KLEINBERG

Hurricane 1928 The Glades

Finding the dead after

TRAGIC AFTERMATH
By dawn on Sept. 17, 1928, a gruesome sight: Recovery workers found bodies everywhere — in the lake, on the roads, in the sawgrass. Bodies soon were so decomposed it was hard to tell blacks from whites. Sue Day, who lives in Lake Park, found this photo in a family album. It was taken by her brother.

The Great Storm of 1928

"Nothing was left then, but to wait and wait dumbly,
like cattle for slaughter. . ."

Soul-sickening ruin

Nothing but piles of wood, animals, people. "Oh, God! Everything!"

By RUTH ELLEN SHIVE CARPENTER
(Aug. 11, 1903 - Oct. 3, 1987)

Ruth Ellen Shive
Carpenter

This account of the Sept. 16, 1928, hurricane was found among Ruth Ellen Shive Carpenter's papers after her death. Ruth was pregnant with her son, Milton, when the storm hit. Milton still lives in Belle Glade.

We left Tulsa, Oklahoma, on a beautiful day, the first of August 1928, little knowing our destination was not Florida, but Tragedy.

I must describe this little town of Pahokee, situated on the shore of Lake Okeechobee. It is a freshwater lake, about 40 miles wide by 70 miles long. All around the lake is a ridge of land, higher than the surrounding land. It is perhaps 150 feet wide. Most of the better homes are built on this ridge, and in this way, the town is strung out around the lake on one side. Just back of this ridge is the low muck land, surrounded by canals. This is where the winter vegetables are raised.

The muck is organic and will burn. It is not unusual to see the ground smoking where some careless person has started a fire. Ground, or muck, that has been burned is always lower by a few inches than it was before. Weeds grow on this land as large as trees, in only a few months' time.

It is a queer, rather beautiful place.

There is a profusion of flowers, tall palms and trees of many kinds, but underneath its seeming placidity, there is a feeling of the jungle ever approaching. A feeling of something ever going on, of millions of small things felt but not seen, and the nights are alive with sound. It is then the alligators busily forage for their food and all sorts of wild animals are found.

Life is very abundant and vivid here . . . There is a saying that once you get muck in your shoes, you will always return, no matter where you go.

The town sleeps through the three months of summer, and it is a dull, torpid place, but the first of September, it becomes feverishly active and remains so the balance of the year.

Now, five weeks after our arrival home, on Sept. 16, 1928, the government issued storm warnings . . .

The wind was blowing strongly on this morning, but my sister and I were taking our time about the housework, and no one was worrying. My husband and brother-in-law were playing checkers . . .

To our astonishment, my father arrived and ordered us all to pack a bag and go to his home or clear to the heart of the town. My father's home and business were on the lakefront, right in the business district, while my sister's home was farther down the lake ridge.

Now, my father is one of the most fearless individuals I have ever known, and many times when the town had been practically evacuated on account of storm warnings, my father would not move a step, and to insist on us taking a sick man (my sister's husband, Duncan) out of bed when he was so weak was unthinkable to us. My father insisted. We threw a few things half-heartedly and rather disgustedly into a bag, only a change apiece. My father and husband carried Duncan wrapped in his blankets and set out for my father's home.

It was about 9 a.m. The wind was blowing strongly, and the lake was stormy and rough as we hurried to the car. We went to my father's home, but after about an hour, he insisted we all stay at the home of my brother-in-law's mother, Mrs. Moran, which was still on the ridge, but farther back from the lake, just across the street from my father's home. My father had built and sold the house himself, and he knew how strongly it was constructed.

A beautiful sand beach stretched out just a hundred feet from my father's back porch, and there was the lake, a magnificent view and a wonderful way to spend a lazy afternoon.

Hour by hour, the wind grew in intensity, and, as other homes were deemed not safe, our little family group became augmented by neighbors and friends. By 2 p.m., there were about 30 of us in the six-room house, and the wind was so strong a person could not stand up outside.

Gradually, the wind grew stronger, and the noise became so deafening that we could only hear one another speak by yelling in the ear. Picture if you can, this group of people all sitting, quietly waiting. As I was the only one who knew shorthand, my ear was laid against the loudspeaker of the radio, and as I would hear a faint voice broadcasting the new location of the storm center, I wrote it on a piece of paper, and it was rapidly handed around the room, until finally we knew we would be near the center ourselves, as the storm had changed its course, and the last faint whisper from the radio faded out.

Nothing was left then, but to wait and wait dumbly, like cattle for slaughter.

The noise became so intense that we all had a peculiar sensation of deafness and tingling in our ears. We watched the barometer fall at a speed that seemed impossible.

About 10 p.m., there was a sudden slacking of the wind, and every face there was hopeful. We believed the worst to be over. We cautiously opened the door and received a few more neighbors who had come in from the muck. My father and mother stepped outside the house to look at their car parked just in front of the door. They had just come in and closed the door behind them when a sudden blast shook the house. It did not come gradually this time, but with fury renewed many times. It was as if all the devils in hell were shouting in our ears. The house shook like a leaf. A gable was taken off the upstairs room and, all at once, we noticed water coming up in the floor, just an inch or two at a time.

I don't believe there was a human being there who was not praying to their God.

I prayed with a new secret fear: the fear of losing the little one close to my heart.

I felt the time was near. My face broke out in an agony of perspiration. I hadn't even the comfort of my husband's presence. Duncan clung to him desperately, begging him not to leave him even if the house went down. I clung to my little gift and prayed. No doctor near. No nothing.

The children, about 10 of them, were all piled on one bed, Duncan on another, and on the only other bed, a neighbor woman who was ill. There was not even chairs for all of us, so that some of us sat on the floor until the water started rising.

I think we all felt the rising water to be the beginning of the end. One side of the room, where my brother-in-law lay, suddenly caved inward as if a giant hand had given it a push and then stopped.

Gradually now, the water stopped rising and the wind began to die down.

By dawn, the wind was still howling, but it seemed like nothing to all of us. I shall never forget my first look at what had happened outside. Such havoc and desolation was soul-sickening. The Everglades is normally a jungle of trees, flowers, grass, weeds or something growing on nearly every inch. There was nothing but ruin. The ground was swept clear of every blade of grass, then huge piles of driftwood, animals, humans. Oh, God! Everything!

My father's house, at first glance, looked all right, but it leaned drunkenly to one side and the back end, toward the lake, was entirely gone. It was off the foundation, and huge blocks big enough for a man to pass through were thrown here and there. The paved street that had separated the two houses was piled high with huge chunks of solid muck. What had caused most of the damage was the veering of the wind when it had stopped suddenly and we thought the storm was over. It had been blowing toward the lake, then it changed and came back, bringing the lake water with it in a solid sheet like a cloudburst out over the ridge, inundating the low muck land behind the ridge. That was when the water came up in our floor, only we didn't know why. We thought perhaps it was the rain pouring through the gable that was blown off. Yet I think each one had a secret fear it might be lake water.

People who were in the low muck land were drowned like rats in a trap. Only a few had miraculously escaped.

"I don't believe there was a human being there who was not praying to their God. I prayed with a new secret fear: the fear of losing the little one close to my heart."

One friend of ours survived by clinging to the back of a dead cow. One mother, with her three little children, found a precarious perch on top of a floating roof and fought all during the night to keep herself and her children hanging on, fighting off the snakes and rats.

Had we stayed in my father's home, probably none of us would have been alive to tell the tale . . .

Even the beautiful beach was an ugly thing of muck, and queer heaps and driftwood. The water, usually a beautiful blue, was now black.

All communication with the outside world was cut off. The paved highways were still covered with water. Only the ridge was dry land. Railroads were washed out.

It looked as if we were on a long, narrow island. Airplanes came at once with food and supplies. By the next day, the town was put under military law. The strong, fierce sun came out and the water began to lower gradually. By the third day, a horrible stench was all over the land. It was painful to even breathe.

Recovery of the bodies was under way. Every man willingly did what he could.

GRIM TASK
The bodies of blacks and whites were separated. Many of the whites were put in coffins and trucked to Woodlawn Cemetery in West Palm Beach for burial. Bodies of blacks were piled up and burned on the spot or moved to a mass grave on Tamarind Avenue. Today, a marker stands at that mass grave, and there is also a marker at the cemetery in Port Mayaca.

HISTORICAL SOCIETY
OF PALM BEACH COUNTY

The bodies were piled in trucks and taken to Palm Beach, but the hot sun and water had rendered most of them unrecognizable after three or four days. My husband saw one truck that would only hold four bodies, they had become so bloated and horrible. Bodies were brought in and piled up just back of what had been the pool hall and burned in great heaps.

An uncle of ours had a home in the suburbs of Palm Beach, and we decided on the fourth day to go there.

It is 40 miles from Pahokee to West Palm Beach, and about half of the highway was water clear up to the horn of the Ford car. There were deep canals on both sides of the road, and with water over everything, you couldn't tell what was canal and what was road.

Finally, Palm Beach stretched before us. It looked like a veritable paradise to us in spite of the fact that here, too, the storm had done much damage. Many of the beautiful hotels and resorts in Palm Beach that were not damaged were thrown open to the public, also many private homes.

The next day, there came a scourge over our world that must have been like the pestilence visited on the Egyptians in Biblical times. Mosquitoes larger than flies were blown in by the storm. Now, we had seen Florida mosquitoes, but nothing like these. They covered the windows in a dark cloud. It was almost impossible to see out.

The Red Cross and Salvation Army fed and clothed everyone. Food was kept in sacks, so much in each sack according to the size of the family.

We went back to Pahokee at the end of three weeks. Coffins and lime still lined the highways, and bodies were being recovered every day, but everywhere the place was in the process of being rebuilt. Stores were opening once again. Roads were being remade. The atmosphere was clean and sweet once again, and the lake its own beautiful blue, and the world seemed at Peace once again.

One of the stores sent word to all the women to come pick out a new fall hat for free. Needless to say, many women received one of the latest, up-to-the-minute, chic fall hats, and I had mine.

Now the town settled down to its usual winter hum of industry, only more so. By November, the town was larger and more beautiful than ever. The climate was perfect: a warm, lovely winter. Birds came in huge flocks every day, until the place seemed a paradise of birds, flowers and sunshine.

Only occasionally would we get a glimpse of something that would bring back the horrible experience of the storm. Once, on our way to Palm Beach, we thought we saw a booted foot sticking up out of the canal. Once when my husband, with a crew of Negroes, was plowing with a tractor, they dug up a skeleton.

My baby was expected the last of January. Christmas day dawned beautiful, warm and clear, and in the early morn on this Holy day, our son, Milton, was born. A beautiful, perfect child. Born on Christ's own birthday. The whole town rejoiced with us.

People I scarcely knew came with gifts and good wishes. A new life was born, and I am sure the rest of that winter will be remembered as one of the happiest of our lives.

Florida is still the land of adventure and pioneering. I am not at all afraid of another hurricane, and isn't it true that all beauty and growth comes through suffering?

I should not want this story to be the means of frightening anyone about living in Florida, but rather an encouragement.

It is a place of honest people, and I think I'd rather live there than any place I know.

"Florida is still the land of adventure and pioneering."

DOWNTOWN WEST PALM BEACH The county courthouse is the columned building on the right (**below**). More than 30,000 buildings were destroyed or damaged in the 1928 storm.

HISTORICAL SOCIETY OF PALM BEACH COUNTY

ACKNOWLEDGMENTS

The entire reporting and photography staffs of
The Palm Beach Post contributed to this book.
In addition to the editors on page 5,
these people made this book possible:

Editor: Edward M. Sears

Assistant managing editor/projects: Bill Greer

Deputy metro editor: Maria Garcia

Assistant metro editors:
Douglas Kalajian   Tom Dubocq
John Bisognano   Bruce Lind
Douglas Zehr   Mary Hladky
Robert P. King   Gary Kane

South County city editor: Price Patton

Assistant state editor: Jane Smith

Treasure Coast editor: Glenn Henderson

Photo editors: Loren Hosack, Jennifer Podis

Treasure Coast photo bureau chief: David Lane

Photo coordinator: Gwyn Surface

Photo lab technicians:
Ray Graham, Tim Stepien, Jonathan Stein

Assistant managing editor/business:
Rick Christie

Assistant business editors:
Paul Bomberger, Greg Stepanich

Assistant managing editor/sports: Tim Burke
Deputy sports editor: Nick Moschella

News editor: Rick Robb
Deputy news editor: Eric Weiss

Suburban editor: Tom Peeling
Assistant suburban editor: Carol Rose

Researchers:
Sammy Alzofon, Krista Pegnetter, Melanie Mena

Color correction: Tim Kelly

Promotion and sales:
The marketing, circulation and advertising
departments of *The Palm Beach Post*

Special thanks to contributors Ray Graham
and Janis Fontaine

WINGIN' IT
Even the birds got sick of the storms. Here, a
pelican tries to dry its wings as wind whips and
waves crash over the seawall along Mobile Bay as
Hurricane Ivan's eye passes over Fairhope, Ala., on
Sept. 16.